GUIDELINES FOR TECHNICAL WRITING FOR LIBRARIANS & INFORMATION PROFESSIONALS

GUIDELINES FOR TECHNICAL WRITING FOR LIBRARIANS & INFORMATION PROFESSIONALS

S. Seetharama

Ess Ess Publications
New Delhi

Guidelines for Technical Writing for Librarians & Information Professionals

ISBN: 978-81-7000-741-8

₹ 425/-

First Published 2015

Published by:
Ess Ess Publications
4831/24, Ansari Road,
Darya Ganj,
New Delhi-110 002.
INDIA
Phones: 23260807, 41563444
Fax: 41563334
E-mail: info@essessreference.com
www.essessreference.com

Cover Design by *Patch Creative Unit*

Printed and bound in India at Salasar Imaging Systems.

Dedicated
to
MY PARENTS

CONTENTS

PREFACE

This book grew out of the need to acquaint the students of Library and Information Science courses and Information Professionals with the fundamental concepts and prevailing practices necessary to understand and use the techniques associated with Technical Writing. The book has been so designed as to function as a basic guide for both potential and actual Technical Writers in the Library and Information Science Profession.

The book has been structured to cover all aspects of Technical Writing. Following an introduction to the subject of Technical Writing, the book highlights the pre-requisites for technical writing, the Basic guidelines and Methodology for the preparation of Technical Communications - including organization and presentation of data / information in abstracts, technical reports, progress reports, state-of- the –art report / review articles, news letters / house journals, audio-visual presentations, etc. A short note on Guidelines for Content Development Strategy for web content development has been given. In addition, it discusses the Editorial process involved in Technical writing and provides a list of useful editorial tools.

The Guidelines given are based on the generalization of experience gained in teaching the subject, as well as in the review of literature and in the preparation of information consolidation products. It is hoped that it would stimulate the librarians and information professionals to master the art of technical writing, a necessary ingredient for information services generation. Perhaps, it would be a useful starting point if Library Schools emphasise on Technical Writing in their academic programmes.

Hyderabad **S. Seetharama**

Chapter A
INTRODUCTION

1. PREAMBLE

Scientists need to be able to communicate with lay people so that their efforts can bear fruit. However, it has been found that Science has become too specialized, and words that mean one thing in one discipline mean something elsewhere. Sometimes, fancy words and / or jargon are used in an effort to be precise, but, in fact precision is lost through confusion. (Alden, 2014).

Consequently, Technical Writing (TW) becomes necessary to make message / information readable, understandable and usable by the target audience concerned. Thus, TW is not creative writing and is markedly different from scientific writing in that it predominantly is written by a scientist for the scientific community. It is more like textbook writing, except that you are required to explain a method, product or process in such a manner that the end users might understand how to perform an operation, expedite a process or use a product or price equipment. In fact, factuality and clarity are of utmost importance, and there is little room for imagination (Lampe, 1970). In other words, the finished product should be readable, clear, concise and understandable, i.e., TW is all about turning a morass of unstructured information into something that people can actually read and use.

Perhaps, you are wondering as to why we should be interested in TW. It is to ensure efficient and effective utilization of the information services and products we as information professionals generate in response to users' needs and thereby facilitate information transfer and communication of ideas. However, most of the scientific and technical literature is followed only by the highly initiated, in spite of the fact that it contains information potentially useful to a wider audience. In other words, the complexity of S & T literature, the language and style adopted seem to act as effective barriers for its wider use. To tackle this anomalous situation, authors have resorted to the technique of TW whose purpose is to make the reader want to read, and digest what has been written.

Perhaps, it will be good and useful for technical writers to apply a kind of communication check-up using the Rudyard Kipling Formula:

"I have six honest friends. They taught me all I know. Their names are: What, Why, When, How, Where, and Who."

The answers to these will help one to plan an effective and appropriate technical communication (Turner, 1974).

WHAT -- is the concept idea, piece of information that one wants to put across? Is it familiar or entirely new?

WHY -- does one need to communicate? (To instruct, To inform? To reprimand?)

WHEN -- is the best time? (Timeliness)

HOW -- can one interest them from the start?

-- Can one best communicate? (Speak/ Write?) Individually or in groups? Direct or through other channels?

-- Are they likely to react?

WHERE -- will it go down best? Situation?

WHO -- are we communicating with? What are the barriers likely to be? Are they the right people?

In other words, every writer should try his best to put adequate efforts to make his writing readable, understandable and usable by the audience(s) concerned. This is easily said than done as "easy reading is difficult to write". For this purpose, one needs three main skills --namely, communication skills, domain knowledge plus computer literacy & word processing skills — which can easily be acquired.

However, it must be said that TW is in its infancy in India as this subject is not taught in formal programmes of Library Schools.

2. DEFINITIONAL ANALYSES

In attempting to discuss the subject of "Technical Writing", it would be helpful to define and delimit the use of the terms and consider some of the resulting implications. This is especially because the meanings of terms depend on the context in which they are used.

2.1. Communication Concept

Communication is the indispensable activity of mankind and begins with birth and ends with the last breath. It is a process of conveying information to influence people for a response. In scope, Communication is so vast and covers language, para-language, body language, signs, symbols and signals of both the animate and the inanimate.

The word 'communication' originates from the Latin root of 'communicare' and 'communis', which means the sharing of information is common. Communication is essentially an art by which a person shares knowledge, feelings, ideas, information, etc, in a way that each gains a common understanding of the meaning, intent and use of the message.

Communication may be defined as:

- act or instance of transmitting;
- imparting or interchange of thoughts, opinions or information by speech, writing, etc;
- something imparted, interchanged or transmitted;
- document or message imparting information, opinion, etc;
- process by which information is exchanged between individuals or groups through a common system of symbols, signs or behavior.

The above statements imply that for information to be transmitted, it has to be first mentally conceived, generated or formed by an individual. Only then can the information be communicated to another. The transmission process could, however, be oral, through visual images or icons, or through written text in any language; or a combination of all these different modes of communication / presentation. As a matter of fact, seventy percent of a man's active life is spent in communication and involves in all activities like reading, writing, listening, speaking, scolding, criticizing, conversing, singing, praising, arguing, instructing, pleading, justifying, loving, hating, comprehending, opposing, imitating, soothing and healing.

Thus, one can see that, generally, TW is a written form of communication and is a mode for transmitting and receiving information.

2.1.1. Objectives of Communication

The objectives of Communication may be stated as:

1. To facilitate effective decision-making;
2. To develop / provide information & understanding required for effective teamwork;

3. To attain higher productivity and higher job satisfaction;
4. To form positive attitudes required for motivation, action, cooperation, coordination, etc; and
5. To help people understand the need for a change and to reduce the resistance to change.

On the other hand, poor communication leads to misunderstanding, insecurity, conflict, wrong interpretation of facts and decision (Seshaiah, 2009).

2.1.2. Communication Process

Communication process is a complex activity consisting of several elements and involves:

1. Source of information;
2. Mode of transmission; and
3. The characteristics and behavior of persons i.e. sender and receiver.

Most communication models employ the concepts of: A sender; A process of encoding into signs or signals or symbols; A message; A channel; A process of decoding; A receiver; Actual or probable effect; and some models incorporate a feedback link.

SENDER → ENCODING → MESSAGE → CHANNEL → DECODING → RECEIVER

FEEDBACK ←

Harold D Lasswell's verbal model, which became known as the Lasswell Formula:

"WHO *says* WHAT *in* WHICH *channel to* WHOM *with* WHAT *effect*" *is given below:*

Lasswell Formula Model

Who	**Says What**	**In Which**	**To Whom**	**With What**
Sender	*Message*	*Channel/ Medium*	*Receiver*	*Effect / Impact*

On the other hand, the mathematical model of communication designed by Claude Shannon and Warren Weaver is as follows:

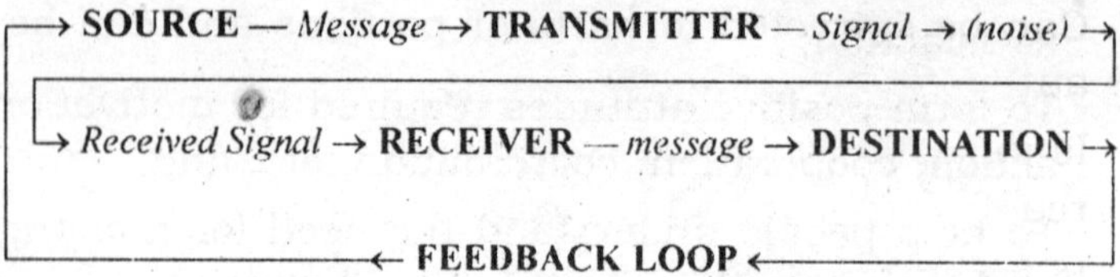

(**Note :** *On its way, the 'signal' can be distorted by interfering 'noise' which means that the signal received by the receiver is not necessarily identical to the one sent by the transmitter*)

The above two models are indicative of the basic components of the communication process.

2.1.3. Forms of Communication

Major types / forms of communication are oral and / or graphic. While oral communication refers to spoken form, graphic refers to written form. One other way of grouping is Verbal and Non-Verbal.

2.1.3.1. Verbal

Oral communication is a method of communication through speech, and has the characteristic of immediacy and personal touch and facilitates immediate interaction between the sender and receiver. When one talks about verbal communication, it would include both oral and written communication.

While written communication does not have the quality of personal touch and immediacy of oral communication, it has an "at a distance" approach and has the characteristics that make it far more effective than oral communication. Some of the perceptible characteristics are:

- Has a formal perspective, suggesting greater authority and trust worthiness.
- Can be a potential record capable of extended life and of being used again and again.

- Can be extracted in parts or portions and can be authentically quoted.
- Is suitable for reading at an individual's own best reading time, speed & pace; and
- Is accessible to a large readership which can lead to the benefit of expert criticism and review by peers.

2.1.3.2. Non-Verbal

Non-verbal communication includes gestures, facial expressions, symbols, signs, signals, pictures, para-language -- sounding like Ah!, Oh!, Ish!, Clap, etc -- and kinesics / body language, eye contact, lip and limbs movement, etc, which makes communication more effective.

2.1.4. Channels of communication

The channels through which the oral or written communication is transmitted cover the **Formal** -- downward, upward, horizontal and lateral --, and **Informal** -- grapevine, exchange of information at casual private and / or public meetings, etc.

While Formal channel is that which is planned, organized and systematic, Informal channel includes all unorganized, unplanned and non-systematic. Formal media include print media and are generally graphic though there are a few kinds which are oral. Some are a combination of these -- lectures, conferences, public addresses, class room teaching, demonstrations, etc.

On the other hand, informal media generally is an oral form of communication and includes tea-talks, travel-talks, and exchanges at casual meetings. "Invisible Colleges" and "Technological Gatekeepers" are two most popular examples of informal media.

2.1.5. Communication Skills

These include personal interaction, speaking, listening, presentation, writing, etc.

2.1.5.1 Definition

In simple terms, communication skill may be defined as a skill wherein one is able to convey information efficiently and effectively.

Other definitions found in literature are:

- as the ability to use language (receptive) and express (expressive) information.
- is the set of skills that enables a person to convey information so that it is received and understood.
- refer to the repertoire of behaviors that serve to convey information for the individual.
- is the ability an individual displays / demonstrates consistently to effectively communicate with clients, colleagues, subordinates, and supervisors in professional manner and in the personal department.
- is generally understood to be the art or technique of persuasion through the use of oral language and written language. (Ronald Allan E. Alfon, 2011).

2.1.6. Technical Communications

Technical communication is a broad field and includes any form of communication that exhibits one or more of the following characteristics:

- Communicating *about technical or specialized topics*, such as computer applications, medical procedures, or environmental regulations.
- Communicating *by using technology*, such as web pages, help files, or social media sites.
- Providing *instructions about how to do something*, regardless of how technical the task is or even if technology is used to create or distribute that communication.

The value that technical communicators deliver is twofold: They make information more useable and accessible to those who need that information.

Society of Technical Communication has given the following examples to illustrate the value of the products technical communicators produce or the services they provide:

- Software instructions help users be more successful on their own, improving how easily those products gain acceptance into the marketplace and reducing costs to support them.
- Medical instructions help patients and care-providers manage a patient's treatment, improving the health of the patient while reducing costs and risks associated with incorrect care.
- Functional specifications and proposals help one group of technical experts communicate effectively with other technical experts, speeding up development cycles, reducing rework caused by misunderstandings, and eliminating risks associated with miscommunication.
- Training programs provide people with new or improved skills, making them more employable and their organizations and products more efficient and safe.
- Well-designed websites make it easier for users to find information, increasing user traffic to and satisfaction with those websites.
- Technical illustrations clarify steps or identify the parts of a product, letting users focus on getting their task done quickly or more accurately.
- Usability studies uncover problems with how products present themselves to users, helping those products become more user friendly.

2.1.6.1. Categorisation of Technical Communications

The following indicates the different types of TCs:

i. Division by function

a) *Proposing what one wants to do* : Proposal for new projects, new methods and materials, research problems, etc;

b) *Recording, setting down details of an action, decision plan or agreement* : Minutes of meetings, Business agreements, transaction;

c) *Reporting, Routine, Periodic* : Organisation reports (annual, half-yearly, etc.),Progress reports;

Inquiry : Surveys, Investigations, Experiments, Tests, Research, etc;

Informing : Information bulletins, Literature reviews, Product descriptions, Process explanations;

d) *Recommending* : Commission Reports;

e) *Suggesting action*: Committee Reports;

f) *Persuading, urging, convince* : Construction bids, Grant applications, An action, Business promotions, etc.

g) *Training / Instruction* : Course materials, Reference guides, System manuals, User manuals.

ii) Division by Type of Documents : Research papers, Professional writings, Instructional materials, Reference materials, Bibliographic tools, Reports, Periodicals, Regulatory & legal documents, Newsletters/ house journals, and such other primary, secondary and tertiary materials;

iii) Division by Readership : Professional peers and colleagues, Business persons, Industrialists, Academicians / Scholars, Bureaucrats / Government Executives, General Public.

2.2. Technical Writing

TW has been defined (Saracevic & Wood, 1981) as a presentation of the essential features of scientific and technical subjects and issues in ways that will make them more understandable and easier to absorb by a wider audience. The aim of it is effective application and wider understanding of scientific and technical advances. TW is then the scaling of scientific and technical information to a given audience. In other words, TW should promote communication by conveying the message concerned. Hence, Factuality & Clarity are most important, and there is little room for imagination (Lampe 1970).

TW, unlike creative writing is not a reality, but, rather the re-creation of what was once a reality, namely, narrating a given job; to meet a social, economic, commercial or techno-economic need. But, the predominant characteristic that distinguishes TW from other expository writing is the reporting of technical data, which necessarily include tabular presentation of data, graphical representation and often photographs.

2.2.1. Characteristics of Technical Writing

Technical Writing invariably:

1. Deals with a subject or topic;
2. Addresses a specific target group;
3. Presents the contents in a language that is oriented towards a target group;
4. Presents the write-up in a form, supported by well designed illustrations and photographs wherever necessary, to draw attention of the target group; and
5. Gets printed attractively.

2.2.2. Principles of Technical Writing

Saracevic & Wood point out that for TW to be effective, it has to adhere to principles of good writing in general, and offer more specific principles including:

- **Accuracy :** make sure that the material selected for TW is accurate in accordance to the most up-to-date information available.
- **Logical Progression :** follow a logical progression of thought.
- **Simplicity :** keep it simple.
- **Word choice :** select words for precise communication, avoid jargon.
- **Style & Grammar :** choose style and grammar for direct & effective communication: not art.
- **Analogy :** make liberal use of examples familiar to the sensitivities of the audience.
- **Sensitivity :** follow any restrictions and adhere to the sensitivities of the assignments, term papers audience.

In addition, the *Handbook of Technical Writing* lists five ethical principles one has to adhere to when writing:

- Avoid language that attempts to evade responsibility.
- Avoid language that could possibly mislead readers.
- Do not de-emphasize or suppress important information.
- Do not emphasize misleading or incorrect information.
- Treat others fairly and respectfully.

2.2.3. Difficulties in Technical Writing

Some of the difficulties in Technical Writing are:

1. Subject matter to deal first and what later;
2. Words to use from a rich variety of English Vocabulary;
3. How to form sentences;
4. Which ideas to be emphasized, omitted or referred to elsewhere, etc.

2.2.4. Technical Writing vis-à-vis Library and Information Professionals

Communication skills, especially written communication skills, are one of the most essential requirements for any professional and information professionals who are basically communicators are no exception. This is because the expanding library & information systems, services and management call for writing skills of the highest quality. There are three writing situations in which TW skills become essential for information professionals.

These pertain to:

1. Individual Professional Development : where writing skill becomes important from the time of joining library schools. Assignments, term papers, project reports, theses, etc require TW skills and such skill would stand an information professional in good stead and help him in ascending to higher positions.

2. Organisational Requirements : Information professionals require different types of writing skills for office administration and management.

3. Professional Services : In the generation of condensation, consolidation & repackaging services and products, L & I professionals need to possess TW skills.

The following presentation indicates the writing situations and target group:

	Writing Situation	Target Groups
1.	Persuasive Writing (Proposals, Promotional materials, etc)	Directors, Top managers, Special Committees, Consumers of products.
2.	Instructional Writing (Course materials, Guides, Manuals)	Students, Teachers, Professional & Technical Personnel, Users
3.	Writings Aiding Decision-making	Heads of • Institutions • Depts of Govt

		• Business & Industrial Undertakings, etc.
4.	Professional Writings (Scientists, Technocrats, Technicians)	Peers, Experts at various levels, Generalists, Non-Professionals, Lay-persons.

2.3. Information Consolidation

UNESCO symposium on Information Analysis and Consolidation in Sri Lanka has given the following definition:

"Information Consolidation Activities are used to define the responsibility exercised by individuals, departments or organizations for evaluating and compressing relevant documents in order to provide definite user groups with reliable and concise new bodies of knowledge. Individuals or groups of individuals performing consolidation activities would constitute Information Consolidation Unit".

Saracevic & Wood (1981) have given the following comprehensive and concise definitions:

"CONSOLIDATED INFORMATION is public knowledge specifically selected, analysed, evaluated, and possibly restructured and repackaged for the purpose of serving some of the immediate decisions, problems, and information needs of a defined clientele or social group, who otherwise may not be able to effectively and efficiently access and use this knowledge as available in the great amounts of documents or in its original form. The criteria for selection, evaluation, restructuring, and repackaging of this knowledge are derived from the potential clientele".

To be precise, "Consolidated Information is a text(s) or messages purposefully structured from existing public knowledge to affect the private knowledge and decisions of individuals who otherwise may not be able to effectively and efficiently access or use this public knowledge from the original documents or in the original structure and form".

On the other hand, the term "Information Analysis" in the context of various activities in the field of information does not connote and denote one and the same idea operationally. While in the context of indexing it is subject analysis and content analysis in abstracting, it is considered to mean critical evaluation when one refers to an Information Analysis Center. In the context of Information Consolidation, however, it is used primarily to denote a non-evaluative information analysis for the purpose of consolidating the results of analysis into a textual presentation in which the isolated units of input undergo substantial structural changes. The product, thus always is factual, non-critical or non-evaluative, but informative in nature. It is a firm, coherent, and unified body of information. This approach is slightly different from the UNESCO and other definitions, wherein, it is emphasized that the most important consolidation function was evaluation and it is this function that distinguishes information consolidation from processes such as, indexing, abstracting, etc.

2.4. Packaging

Packaging of information is a physical recording, arrangement and presentation of information on a given medium and in a given form. In packaging, two aspects can be distinguished:

1. Package Media : The reference here is to physical medium in which the information is displayed or presented to users. For example, Print, Audio-visuals, demonstrations, inter-personal contact, etc.

2. Package Formats : The arrangement, shape, and lay-out of information in a given production on a given medium.

At this juncture, it may be emphasized that while packaging cannot improve information content, it can enhance its use, assimilation and recall.

2.5. Repackaging

Repackaging of information is rearrangement of physical media and /or forms in which information has been presented, or presentation in a given medium and form of restructured information, which is tailored to the requirements of a specific clientele. In other words, repackaging of information refers to the presentation of information in more understandable, readable, acceptable and usable forms.

The aim of repackaging is to enhance the acceptance and use of information products, and the assimilation & recall of their contents.

The functions served by repackaging of information are:

- as a saving tool (time saver)
- as a selective and systematic sorter of useful information
- as a means for more extensive information transmission and delivery
- as a translation tool
- as an opportunity for the practical application of research results
- as a means for the prompt delivery of relevant information. (Saracevic and Wood, 1981).

3. READER WRITER RELATIONSHIPS

The relationship between reader-writer is said to be intertwined. On his side, a writer can write without a reader, but his words, would not mean anything if they (words) remain unread and unexplored. On the contrary, a reader cannot exist without a writer. Thus, the relationship is one of symbiosis - - a relationship of mutual benefit or dependence.

In scientific communication, there is a pattern of flow of information from generator (Writer) to the (Reader) wherein a quality mechanism exists, there is a built-in reader-writer relationship. Similarly, in TW situations, there is also a reader-writer relationship although not so explicit. In addition, reader response meaning the behavior and reaction of uses --, and feedback which refers the way they express their reactions to the technical document is a normal feature in scientific communication.

A similar situation prevails though not so explicit in technical writing situations. Reader response and feedback in TW can be gauged indirectly through book reviews, review articles, annual reviews, citation analysis, etc. In the context of instructional documents like course materials, operation manuals, literature guides, etc., one can assess users' reactions through:

1. Structured questionnaires covering the area of content readability, structure, style of presentation, understandability, usability, etc;
2. Performance test in case of operation manuals, guides; and / or
3. Protocol analysis which allows the writer to record readers' responses while they read through a document. The responses recorded may pertain to content, structure, style, use of jargon, sequence of presentation, completeness, comprehensiveness, preciseness, clarity, etc.

On the other hand, in investigation reports, the terms of reference form the basis in assessing users' reactions directly or indirectly through meetings, which highlight the features and qualities of information services and products. These promotional materials ensure a means of communication and thereby a reader-writer relationship.

At this juncture, stating the following tips for good writing that ensures a healthy reader-writer relationship will be in order.

- To be a good writer, one must form a clear mental picture of the reader and speak to him in an appropriate tone as a real person in a language that he / she understands.
- Tell the reader what you (writer) want him to know and all the reader wants to know -- background information, related readings, source information, etc.
- Provide as much context as possible in your writings -- historical and cultural, time context, etc.
- Anticipate and answer your readers' questions.
- Adopt different modes of presentation - -textual and non-textual -- of information.
- Take your reader on an interesting and well considered journey.
- Respect the time your reader chooses to spend with your writing.

4. LANGUAGE AS A MEDIUM OF COMMUNICATION

Earlier, it was emphasized that the language used in technical writing should be one that facilitates communication of information. It has been stated that language(s) arose because human beings needed a greater degree of cooperation with each other in order to survive, and this cooperation required an effective system of communication. In addition, language is used to communicate feelings and emotions known as phatic communication and it is a device to establish and maintain social contact. Language also is said to possess a "clue- bearing" role whereby it indirectly conveys information about the speaker, writer, like the region from where he / she comes, his / her social status, occupation, etc. Lastly, it is also used for aesthetic reasons for purposes like writing a poem or novel or drama, etc.

Irrespective of the fact that one and the same language may have dialectical differences due to regional separation, socio-economic status, caste, ethnic group, sex, etc, structural and functional differences, differences in spoken and written communication, it should be remembered that language is a medium for communication of thought.

5. READABILITY AND TEXT

Readability is concerned with the problem of matching between the reader and text. While an accomplished reader is likely to be bored by simple repetitive texts, a poor reader will become discouraged by texts which he / she finds too difficult to read fluently. This is likely to happen when the text is poorly printed, contains complex sentence structures, long words or too much material containing entirely new ideas.

Readability, thus, may be defined as reading ease, especially, as it results from a writing style. Research, however, has shown that easy reading text improves comprehension, retention, reading speed and persistence.

5.1. Factors Affecting Readability

Ease of reading, in its turn, is the result of interaction between the text and reader. The factors affecting readability in relation to the reader are:

1. Prior knowledge;
2. Reading skill;
3. Interest; and
4. Motivation.

On the other hand, in the text, factors affecting readability are:

- Content;
- Structure;
- Design; and
- Style of presentation.

Content : Of course is most important as the stuff

that goes into the technical communication. The first and basic rule is that quality of ideas and expression is most important.

Structure : All written material, whether it is a one-page write-up, lecture summary or a long report, must follow a structure. The most basic structure is: Introduction, Body, Conclusion - - while the Introduction lays out briefly what the technical communication will do, with a few pointers that give the reader a sense of what to expect, the Body provides the main argument(s) along with the evidence to support it. The Conclusion for its part summarises the main argument(s), leaving the reader with something to think about. As one is structuring the technical communication, descriptive sub-headings, section headings should be used as these stand out and serve as signposts to the reader, and break the monotony of running text.

Design : It includes the medium, lay-out, illustrations, reading and navigational column width, text colour, background contrast and white space make text easy to reader.

Style of Presentation : Style adopted should facilitate understanding and be pleasing to the reader. It should focus on accuracy, brevity, clarity and readability and relates to terminology, grammar, punctuation and mechanics.

According to linguistic experts, Readability is a score produced by Readability Formula, which is usually calibrated against a more labour-intensive Readability survey.

Some of the common Readability Formulae are:

- Automated Readability Index
- Coleman-Liau
- Flesch Reading Ease
- Flesch-Kincaid Grade Level

- Gunning Fog Index
- SMOG

The formulae are widely used to match texts with the reading level of the audience.

On the other hand, Publishers use Readability Formulae to assess the reading level of the text. They also use word-frequency lists - -the frequency of a word is a good indication of its ease of use. Text leveling, another technique used is a subjective evaluation of a text, based on training and experience, is another important adjunct to using a formula.

In addition, writers, editors and publishers also often make intuitive assessments of readability based on experience, insight into their target audience, and knowledge of a number of rules of thumb, which are often derived from assessing a number of readability survey results. It has been observed that there are some computer programmes for measuring readability of text on the World Wide Web and some of these are specifically designed to measure the readability of web pages. (Wikipedia, the free encyclopedia. enwikipedia.org/wiki/ readability)

6. ABERRATIONS IN TECHNICAL WRITING

While it is absolutely necessary to have familiarity with knowledge & skills for TW as well as its standards, norms and conventions, it is equally important to be aware of the aberrations that are likely to occur in TW. Aberrations in text occur not only from internal shortcomings like insufficient or inaccurate content, weak organization of information, indifferent or careless use of language, poor documentation and unnecessary, irrelevant or inaccurate visuals, but also due to deficiency in communicating its purpose and manage the reader. In other words, a text may be well written and complete in coverage, but ineffective if the needs of the reader are left unsatisfied.

Aberrations can occur due to factors like Content Readability and Technical style of presentation.

6.1. Content Readability

Content readability, in its turn, is ensured by accuracy, preciseness and completeness of the information contained in the text. For this purpose, the writer should be sufficiently familiar with the subject(s) concerned, the possible primary and secondary sources of information to be consulted, the appropriateness of content (information) vis-*à-vis* the reader. Once the relevant information is collected, it needs to be organized and presented as a text that can be easily read, understood and assimilated by the readers concerned. The organization adopted may be logical, chronological, psychological or topical depending on users' needs. In its turn, organization of data / information in a text brings clarity. Hence, in TW, principles of organization and expository writing (wherein, any body of written information should have a beginning or introduction, middle where the theme is fully developed, followed by the end which summarizes the conclusions reached and makes recommendations) need to be adopted. Besides, organization adopted is dependent on the formats of the technical communication concerned.

Further, content readability and readability is enhanced by non-textual modes of presentation like tables, graphs, charts, flow-charts, diagrams, drawings, photographs, etc especially when ideas and concepts in a text are complex and too abstract for the reader. However, these visuals are liable for different types of errors pertaining to the labels and captions given. Reliability also requires sufficient documentation of the information sources used and the formats of documentation adopted is dependent on the source(s) concerned.

In other words, for content reliability one must ensure that no errors creep in:

1. Accuracy, precision, and completeness of information in the text;
2. Organizational & expository writing principles and formats followed;
3. Appropriateness, accuracy and completeness of visuals used; and
4. Acknowledgement of sources of information used.

6.2. Technical Style

Style adopted in TW should facilitate understanding and be pleasing to the reader. This is bound to be variable depending on the audience concerned. Style, generally, should focus on accuracy, brevity, clarity and readability and relates to terminology, grammar, punctuation and mechanics.

Some of the criteria / rules to be followed in relation to style of presentation in TW are given below:

- Use language that ensures precision, brevity, clarity and readability.
- Avoid jargon or use it sparingly.
- Use simple, direct, specific and concrete words instead of general, abstract words and phrases.
- Avoid noun strings, redundant terms and negative phrasing (prefer positive phrasing as it makes the text more readable and direct.)
- In relation to grammatical structures - - use active voice instead of passive voice as far as possible; avoid dangling modifiers and abstract nouns; subject and verb should agree in number and person i.e. Singular nouns with plural verbs and vice versa should not be used;
- Punctuation marks used should be consistent and appropriate; and

- Convention in relation to use of abbreviations, numerals, capitals, quotation marks in documentation, underlining, italics, headings and sub-headings are to be followed.

7. ROLE OF TECHNICAL WRITER

Now, let's look at the functions that a Technical Writer has to do, over and above researching and analyzing the subject matter:

Audience Analysis

- The information fits the purpose of its intended readership
- The tone of writing is correct.

Conceptual Analysis

- Ideas are presented logically and completely
- All assumptions are clearly supported or implied by audience Knowledge

Conciseness

- Superfluous information is removed
- Reference & Instructional information are separated from each other
- Where appropriate, material is 'chunked' to optimize it for easy Comprehension

Structure

- The relationship between headings and body information is Clear

Terminology

- All terms are consistent, appropriate and covered by the glossary
- Abbreviations are avoided where possible

Formatting & Lay-out

- Conforms to a pre-defined standard

Illustrations

- Are of appropriate size and quality for selected media
- All annotations are sensibly placed and refer to terminology used throughout the rest of the document

Language Mechanics

- Grammar, spelling, capitalization, punctuation and sentence structures are correct and consistent
- Instructional information is in the active voice

Language Style

- Pre-defined house style is followed
- Tone is consistent and the flow of text ensures ready Comprehension

Cross-references

To be used are correct, appropriate and complete, but not overused.

Chapter B

PRE-REQUISITES FOR TECHNICAL WRITING

1. INTRODUCTION

Until recently, the technique of TW was practiced by subject specialists who were responsible for the generation of publications both within and outside the organization. But, with advances made in Library & Information Science, information professionals have started playing a dominant role in the preparation quality. However, for them to play a useful role in TW, the following pre-requisites become absolutely necessary:

1. Communication Skills;
2. Domain knowledge;
3. Familiarity with users' requirements / needs on the subjects of their pursuit;
4. Computer literacy and Word processing skills; and
5. Knowledge of:
    - Helpful methods of presentation of ideas;
    - Modes of Presentation;
    - Visual Communication techniques; and
    - General Rules of Grammar, Style, etc of the language.

2. COMMUNICATION SKILLS

Communication skills enable one to efficiently and effectively convey information to the target audience so that the information received is readable, understandable and usable. This ability would be very useful and helpful in one's professional as well as personal life.

2.1. Types of Communication Skills

Some consider the top 10 Communication Skills to be Listening, Non-verbal communication, Clarity & Concision, Friendliness, Confidence, Empathy, Open-mindedness, Respect, Feedback, and Picking the right medium.

Generally, the types of Communication Skills are classified as Verbal and Non-verbal, Technological and Non-technological, Mediated and Non-mediated, Participatory and Non-participatory.

However, the commonly known types of communication skills are:

Intra-personal communication skills : This implies individual reflection, contemplation and meditation. One example of this is transcendental meditation. According to the experts, this type of communication encompasses communicating with the divine and with spirits in the form of prayers and rites and rituals.

Interpersonal communication skills : This is direct, face-to-face communication that occurs between two persons. It is essentially a dialogue or a conversation between two or more people. It is personal, direct, as well as intimate and permits maximum interaction through words and gestures. Interpersonal communications maybe:

Focused Interactions : This primarily results from an actual encounter between two persons. This implies that the two persons involved are completely aware of the communication happening between them.

Unfocused interactions : This occurs when one simply observes or listens to persons with whom one is not conversing. This usually occurs at stations and bus stops, as well as on the street, at restaurants, etc.

Non verbal communication skills : This includes aspects such as body language, gestures, facial expressions, eye contact, etc., which also become a part of the communicating process; as well as the written and typed modes of communications.

Mass communication : This is generally identified with tools of modern mass media including social media marketing, which includes: books, the press, cinema, television, radio, etc. It is a means of conveying messages to an entire populace.

No matter what the different types of communication skills are, communicating is an ever-continuing process that is going on all the time. It is as important to human life as is day-to-day existence.

(***Source:*** www.communicationskills.co.in)

Essentially, for Technical Writing, one needs:

- **Interpersonal communication skills** in face-to-face communication (between technical writer & audience / readers) wherein verbal and non-verbal communication skills come to play;
- **Interviewing &Listening skills :** There is a need to know how to ask questions. Strong listening skills to capture information necessary and to know which follow-up questions need to be answered. (One must distinguish between hearing and listening).
- **Presentation skills**; and
- **Writing skills :** to write in a clear and concise manner so as to be able to effectively convey information appropriately for a variety of audiences as poor written communication can be frustrating

for the reader and potentially damaging for the author. It should be remembered that writing skills include spelling, grammar, structure, and formal & informal writing styles.

- **Usability & Testing skills :** for validation of the technical document prepared.

3. DOMAIN KNOWLEDGE

It is possible to get an understanding of the highways and byways of the subject(s) through:

- Systematic study of subjects through documents using a proforma for collection of information.
- Use of a scheme(s) of classification.
- Design and development of a scheme(s) of classification.
- Consultation with Subject Specialists.
- Attending and participation in lectures and discussion of Subject Specialists

The more intensive the knowledge he has of the subjects, more helpful it will be in making technical communications productive. Obviously, such knowledge of all the subjects cannot be acquired by any one person. A person with a basic background in a subject can build upon it through experience - - such as, doing research and / or teaching the subject. While intensive specialization is not necessary for information professionals, a broad perspective of relevant subjects is necessary so as to serve the clientele concerned. (Seetharama, 1997).

3.1. Specific Implication

In general, the kind of knowledge that information professional should gain about subjects, may be specified as follows:

- Scope of the subject(s) as a whole;

- Its main branches and subdivisions within each branch and the scope of each of them;
- The ideas generally falling in relation to the subject and their grouping in relation to the branches and subdivisions within each branch of the subject;
- The interrelation between the component ideas and the relevant characteristics on the basis of which they can be grouped;
- The landmarks in the evolution of the subject: the important contributors and their respective contributions;
- The state of the art and trend in each of the branches of the subject(s);
- Inter-relation of a subject with other subjects; and
- The technical terminology of each of the different subjects.

4. INFORMATION REQUIREMENTS OF USERS

On the other hand, the study of information needs of potential and /or actual users involves the Study of Institutional characteristics, its projects and needs; Study of potential User groups on the basis of questionnaires, interviews, and a Study of organizational and functional charts, etc; and Study of individual user's interest by adopting different methods of collection of information about the informational requirements of the actual and potential users. *(For details, refer to Seetharama: Information Consolidation & Repackaging. 1997).*

Studies such as those mentioned above may have to be done periodically in order to update the knowledge of users' interests, institutional programmes, etc. There should be a built-in mechanism in the institution for this purpose, as users' information needs are not static, but are dynamic.

5. COMPUTER LITERACY & WORD-PROCESSING SKILLS

Computer literacy is defined as the knowledge and ability to utilize computers and related technology efficiently, with a range of skills covering levels from elementary use to programming and advanced problem solving. Computer literacy can also refer to the comfort level someone has with using programs and other applications that are associated with computers. (en.wikipedia.org/wiki)

Word processing, on the other hand, is the work or skill of producing printed documents using a computer.

As the ultimate product in TW is a printed document, it is necessary that one possesses computer literacy and word processing skills.

6. GUIDING PRINCIPLES FOR ARRANGEMENT OF IDEAS IN A HELPFUL SEQUENCE

In the preparation of Technical Communications, the major items of work are:

1. "Authorial" work ; and
2. Physical production of the document.

In this section, only authorial work is discussed.

Authorial Work

"Authorial" work consists of work in three planes - - namely, Idea Plane, Verbal Plane and Notational / Structural Plane.

Idea Plane

Work in the Idea Plane essentially consists of:

a) Perusing the relevant documents on the subject and collection of ideas for consideration of inclusion in the technical communication;

b) Selection from the ideas collected;

c) Organising the selected ideas in a helpful sequence.

Verbal Plane

Work in the Verbal Plane essentially consists of:

a) Expressing the ideas in a language for the purpose of communication; and

b) Ensuring directness, transparency, and unequivocalness of expression.

Structural Plane

Work in the Structural Plane essentially consists of:

a) Preserving the helpful sequence of ideas arrived at in the Idea Plane; and for this purpose,

b) Formulation and use of a notational system in structuring the text.

6.1. Idea Plane

Guiding Principles for arrangement of ideas in technical communications have already been formulated.

The following are the Principles:

a) Principle of Unity of Ideas;

b) Principles for Sequence of Ideas:

 1) Principle of Filiatory Sequence.

 2) Principle of Decreasing Extension; or, if more appropriate to the context

 3) Principle of Increasing Intension;

 4) Principle of Time Sequence:

  i. Principle of Later-in-Time; or if more appropriate to the context

  ii. Principle of Earlier-in-Time;

 5) Principle of Developmental Sequence;

 6) Principle of Evolutionary Sequence:

  i. Principle of Later-in-Evolution; or if more appropriate to the context

  ii. Principle of Earlier-in-Evolution;

7) Poorva-Paksha Siddhanta Principle; Statement of a principle and the establishment of a more suitable one;
8) Principle of Spatial Contiguity (any one of the following six principles, most appropriate to the context):
   i. For entities in Vertical Sequence;
   ii. For entities in Horizontal Sequence;
   iii. For entities along Circular Line
   iv. For entities along radial Line;
   v. Principle of Away-from-position;
   vi. Principle of Spatial Contiguity;
9) Principle of Quantitative Measure;
10) Principle of Increasing Complexity;
11) Principle of Concreteness *Vs* Abstractness;
12) Principle of Traditional Sequence;
13) Principle of Literary Warrant;
14) Principle of Alphabetical Sequence.

6.2. Verbal Plane

Guiding Principles for the Verbal Plane have already been formulated. These include:

1. Respecting Unity of Ideas;
2. Principle of Simple Sentence;
3. Principle of Transparency;
4. Avoidance of Tautology;
5. One-to-one correspondence between 'Term' and 'Idea'
   i. Avoidance of Homonym;
   ii. Avoidance of Synonym;
6. Principle of Technical Terminology;
7. Consistency in Punctuation;

8. Principle of use of New Term; and
9. Principle of Elegance.

As knowledge of Guiding Principles for arrangement of ideas in the Idea, Verbal Planes and their application is familiar to the information professionals, these are not discussed here. (A reader requiring more information may find it in Ranganathan's Prolegomena to Library Classification)

Chapter C
Guidelines For Technical Writing

1. INTRODUCTION

In the preparation of Technical Communications -- namely, proposals, technical reports, state-of-the-art reports, etc., the emphasis is on the target group concerned on the one hand, and the best available information on the topic on the other. On the operational level, various stages / steps can be identified involving different activities / procedures. However, the technical communications -- the characteristics and types -- to be prepared would be related to the information needs in different stages of the diffusion process. As is known, diffusion can be portioned into five stages or phases -- namely, awareness, interest / knowledge, attitude formation, trial / demonstration, and adoption / confirmation -- with the information requirements becoming more specific and thus technical communications tend to become more specific and even personal. In turn, the effort and cost involved would increase correspondingly. The following Table C1 provides examples of technical communications as they are appropriate for a number of different types of academic / research organizations and as they relate to the five stages in the diffusion process (Saracevic & Wood, 1981):

Table C1: Examples of Technical Communications by Diffusion Stages.

Awareness Stage

Journal notices, Short announcements in professional newsletters, Acquisition

Interest/ Knowledge Stage

State-of-the-art reports, Critical reviews, Synthesized research results, Referral, Journal articles, Text excerpts, Critical data, and Data survey.

Attitude formation Stage

Synthesis of critical reports & evaluations, Comparative results, Resource requirements of information.

Trial / Decision Stage

Research data, Training & Education materials, Know-why information, Experiences at other institutions.

Adoption / Confirmation Stage

Continuing synthesized SDI reports, Selective Current Contents, Refinements in advances, Critical data.

2. BASIC GUIDELINES ON TECHNICAL WRITING

James R. Wilson has provided an aid to students who seek to improve the readability of their reports, theses, dissertations, conference papers, and archival journal articles. This note summarizes some useful guidelines on technical writing.

I. ORGANIZING THE PAPER (what to do before beginning to write)

   A. Analyze the situation, that is, the problem, the solution, and the target audience.

   B. Make a detailed outline, and use it as the basis

for the formal presentation of your work.

II. WRITING THE PAPER

A. Prepare an abstract that is concise, complete in itself, and intelligible to a general reader in the relevant field. The abstract should not exceed 200 words, and it should not contain any references or mathematical symbols.

B. Write the rest of the paper as though you were talking to a group of interested colleagues about your work.

C. For each table, compose a caption that briefly summarizes the content of the table. Comment explicitly in the text on the significance of the numbers in the table; do not force the reader to guess at your conclusions.

D. For each figure, compose a caption (or legend) that explains every detail in the figure -- every curve, point, and symbol.

E. Revise and rewrite until the truth and clarity of every sentence are unquestionable.

F. Prepare a complete and accurate set of references that gives adequate credit to the prior work.

G. To ensure that a thesis or dissertation conforms to the university's requirements for format and organization,

III. ACHIEVING A NATURAL AND EFFECTIVE WRITING STYLE.

Another approach (Tech Prose) suggested includes:

1. Getting Started

1.1 Determine the Purpose and Use

1.2 Identify the Audience and What They Need

1.3 Determine the Level of Detail

1.4 Organize the Data

1.5 When You Work with a Team

1.6 Deadlines

2. Your Role as a Writer

2.1 Employ Ethical Principles

2.2 Give Credit where Due

3. Style & Usage

3.1 Elements of Style

3.1.1 Present Tense, Active Voice

3.1.2 Simple Sentences

3.1.3 Gender Neutrality

3.1.4 Abbreviations and Acronyms

3.1.5 Lists

3.1.6 Sidebars

3.1.7 Figures and Tables

3.1.8 Cross-references

3.1.9 Footnotes and Endnotes

3.1.10 Redundancy

3.1.11 Final Edit

3.2 Wasted Words and Phrases to Avoid

According to Rajashri Rajan (2003), TW is all about creating a usable piece of information or turning a morass of unstructured information into something that people can actually read and use. In other words, the user manuals and the online help systems that one sees integrated with any piece of software that one uses is the result of the efforts of a technical writer. However, TW need not necessarily be associated with the IT industry alone. She discusses below the process involved in TW; some of the types of deliverables; the tools used on the job; and the predominant rules one needs to follow.

TW like any other process can be summarized under the following categories:

Know your audience : collect as much information as you can on what does the reader require the information for, what is their current knowledge base, what are the reader's biases, if any,etc.

Gather information : contact SMEs (Subject Matter Experts) and gather as much information as possible on the subject you are preparing the document on. It is important to keep a free mind and not monitor the order of your thoughts. Make it a point to note down all the information and not attempt to group or arrange information at this point. It is advisable not to restrict yourself to the format or grammar of the ideas.

Categorize and organize information : categorize the information into relevant topics. To begin with, create random topic headings, delete the irrelevant information, group related topics and then arrange the topic decide on the order you are planning to arrange the topics — for example, chronological, general to specific, spatial, etc.

Design document : design the document before you actually write the content. A suitable template needs to be selected or designed and used for writing the document. Some of the points to be borne in mind while designing a template are discussed later.

Write document : now, get down to write the document. It is important to write section by section and not to revise as you write, as you may tend to lose the flow of writing if you do so. Use style guides so as to follow the rules of writing.

Revise document : once you are through with the writing, read through and revise the document once again. The style guide that you have used for reference while writing can serve as a checklist. You may also have an additional checklist that is specific to your project or the client you are writing for.

Document deliverables, on the other hand, include: software manuals, online help systems, administration guides, application notes / briefs, design guides, design specifications, functional specifications, installation guides, quick beginning guides, hardware manuals, reference manuals / books, and so on.

The Technical writer's toolkit includes a set of software for quality documentation -- RoboHELP, Adobe Frame maker, Adobe acrobat, Microsoft office Suite, etc --, Style guides like Microsoft Manual of Style for Technical Publications, and if you are not influenced by the 'Gates' monopoly, the Chicago Style Guide serves the purpose just as much!

The Pre'Scribe'd Rules (prescribed rules).

While writing a document:

- Use an abstract, Table of Contents or other front matter elements
- Use vertical lists
- Use enumerated lists
- Use visual elements
- Write shorter sentences
- Use headers and footers
- Paragraph more frequently
- Break information into sections / chapters
- Use more white space
- Use two-column format, Use right ragged text instead of justified text,
- While writing a document, write section by section
- Write only, and not revise as you write
- Be yourself. Don't mime the style of 'experts'
- Use simple words / sentences
- Keep sentences short
- Repeat nouns to avoid ambiguity
- Don't be afraid to fail

While revising a document

- Use simple words / avoid euphemisms
- Use short sentences
- Use bias-free communication
- Use active voice extensively
- Use passive voice appropriately
- Write numbers in narrative appropriately
- Write abbreviations appropriately
- Maintain parallel construction
- Keep text precise and concise
- Keep to the subject
- Check spellings
- Check capitalization and punctuation
- Avoid low information content words / redundant words
- Avoid vague words / abstract expressions
- Avoid negative constructions. Use only when necessary
- Avoid inverted order.

3. METHODOLOGY FOR TECHNICAL COMMUNICATIONS WRITING

The Methodology to be followed for Technical Communication preparation is discussed presently. The principles governing and the processes involved would generally be applicable to all types of technical communications. As mentioned in the previous chapter (vide sec.6), the work involved can be conveniently demarcated into three planes of work: Idea Plane, Verbal Plane, and Structural Plane.

The work in the Idea Plane involves the following steps:

1. Recognition of the need for a Technical Communication (TC);

2. Team formation for design and development of TC;
3. Objective Formulation & Target Group Identification
4. Subject Scope of TC
5. Collection of Information (pertaining to the subject(s) to be covered by the TC) and involves:
   - Identification of information sources.
   - Selection of information from among the primary and secondary sources of information.
   - Preparation of abstracts from selected documents /sources.
6. Appraisal of information collected;
7. Scope re-definition, if necessary;
8. Draft text preparation involves:
   - Document Plan.
   - Arrangement of ideas.
   - Integration of information involving;
     - Application of Logical / Classificatory principles.
     - Application of Visual Communication techniques, if any.
     - Adopting different Modes of Presentation.
     - Following general rules of grammar, style, etc involved in writing.
     - Orientation, if any, to the needs and levels of users.
9. Testing of Draft Text and involves
   - Review and Assessment
   - Field Testing.
10. Finalisation

While the work in the Verbal Plane involves the use of homonym-free, synonym-free terminology and other factors associated with the expression of ideas in a language for the purpose of communication, work in the

Structural Plane consists of preserving the helpful sequence of ideas arrived at in Idea Plane and for this purpose the formulation and use of a notational system in structuring the text.

3.1. Recognition of the need for a Technical Communication (TC)

The recognition may occur due to any of the possibilities listed below:

a) The Library & Information Science personnel may recognize due to their interaction with the user groups and / or by Brainstorming.

b) The potential users -- beneficiaries and intermediaries -- may themselves express a need and, perhaps, demand a TC.

c) The crystallization of a new discipline or an interdisciplinary area may call for the production of TCs periodically to help the users. Similarly, a new mission within an older activity raises the need.

d) National policies and priorities may require the application / adaptation / adoption of certain technologies necessitating the need for TCs to suit the differential requirements of the different user groups.

3.2. Team Formation

This involves identification and formation of a team of workers / specialists for the design and development of a TC. The directories and the formal / informal contacts of library personnel and information professionals would facilitate this process of identification leading to team formation.

3.3. Objective Formulation & Target Group Identification

The writer / author has to decide for what purpose he is writing -- As a writer, is he trying to introduce a new idea, or instructing, or reporting findings, or recommending / persuading a particular course of action, or facilitating decision-making? This would necessarily mean identification / specification of the target group.

In this context, information on the following attributes of the reader could be helpful:

a) Specific purpose for which the TC is needed. Is it for decision-making, extension activity, research activity, etc?

b) Intellectual standard of the target group.

c) Reader's preference for a particular type of TC and type of presentation of information, etc.

In other words, this step should result in the identification of the reader -- as to whether he is a researcher, managerial personnel, extension worker or general reader.

3.4. Subject Scope of TC

Based on the TC required by the target group, the subject scope should be determined. This involves the precise specification of the subject and would consist of recognizing each of the component ideas and the correct degree of interrelation among them, and expressing the totality of the subject uniquely and coextensively. In addition, key information need areas, depth and period of coverage have to be specified. Further, the type of TC to be prepared needs to be specified.

The specialist groups involved in his step would naturally be those who benefit from the information contained in the TC - - beneficiaries /end users; library & information science personnel; and the author of the TC. At this stage, the Subject specialists / Peer groups who should be consulted in the preparation of the TC have to be identified.

Note *: This is an important step. If communication between the groups is not effective, usefulness of the TC is likely to* *be lessened and the time and effort put on it would then be a waste. It would be a good policy to put down in writing the precise scope and get a confirmation from the user & target groups concerned.*

In determining the subject scope of the TC, the information needs of the potential users needs to be studied. This is quite easy in the case of users belonging to a particular discipline or disciplines. But, in the case of planners and decision makers, it is rendered difficult as the information needs are not structured according to 'subjects' in the usual sense of the term, but are essentially function or task-oriented. That is, the TC would be used in performing certain tasks /functions involving state-of-the-situation analysis, extrapolation and forecasting future needs, contingent decision making and problem solving, anticipating consequences of action, etc.

While the user needs approach may be difficult, if not impracticable to implement and /or may not yield the desired results when the target groups are planners and decision makers, an alternative approach is to use a *Model or Reference Framework* according to which planning & decision making programmes can be conceptually analysed into different phases, sub-tasks, activities, etc., and the information support needed identified for each of these. When such an analysis is done, it would be possible to select /extract relevant data and information from different sources /databases and repackage them into relevant subsets for convenient use at the different stages of planning work. (For details regarding Reference Framework, see *Neelameghan & Seetharama, 1975).*

3.5. Collection of Information

In order to prepare a TC, it is a necessary prerequisite to collect information on the topic selected. This necessarily implies the selection and acquisition of information sources keeping in mind the ideas / concepts going with the precise subject determined in the earlier step. This, in turn, would involve:

Identification of Information Sources

- Reference books and directories, indexing and abstracting sources and services, SDI, new titles information, invisible college, etc.

Selection of information from among the primary and secondary sources of information.

- Identification of selectors
- Formulation or identification of criteria for selection and evaluation of information and information sources *vis-à-vis* intrinsic value and user appropriateness
- Preparation of informative abstracts or extracts of relevant documentary and non-documentary sources.

The L & I personnel play a dominant role in this step. After initial scanning of secondary periodicals, they prepare a priority list or core list of periodicals / documents to be scanned. This is followed by preparation of informative abstracts of relevant documents using certain guidelines for abstracting / condensation. (Gopalakrishnan, 1981, Seetharama, 1978, 1997)

3.5.1. Abstracting

As abstracts facilitate in the preparation of subject bibliographies, state-of-the-art-reports, trend reports and other TCs, they need to be acceptable and adequate from the point of the users. Borko and Chatman have specified that an adequate abstract of a document must contain information on:

a) Purpose of writing the document;

b) Methodology / Procedure followed with a statement of experimental techniques used;

c) Results, i.e., a statement of the findings; and

d) Conclusions, i.e., a statement dealing with the interpretations or significance of the results.

In addition to the above, in certain specialized fields, an abstract should contain specialized information. For example, in the field of medicine, an abstract should contain details of diagnosis and treatment, drug dosages, etc, where applicable.

The different steps involved in abstracting are:

i. Selection of Material for abstracting;
ii. Identification of Focus;
iii. Content Analysis;
iv. Assessment;
v. Synthesis;
vi. Checking; and
vii. Editing.

3.5.1.1. Selection of material for abstracting

Some of the Do's and Don'ts in selecting material are given below:

	Do's	**Don'ts**
i.	Select relevant material/documents keeping in view the interest of the users.	Do not select material/ documents which are irrelevant to the users.
ii.	Select significant material.	Avoid trivial material.
iii.	Select original & novel contributions.	Avoid repetitious material.
iv.	Select documents	Avoid selection of documents
	• Published from reputable sources	• Published in sources of unknown or questionable reputation.
	• Less accessible	• Readily accessible
	• Published in sources a service has Committed itself to cover by policy.	• Published in sources usually excluded by policy
v.	Select final reports & other reports which are adequately supported by data or evidence.	Do not select preliminary progress reports inadequately supported by data or evidence.

3.5.1.2. Identification of Focus

The identification of the focus (Central theme) of the subject embodied in the document is the first step in the preparation of the abstract. Identify the primary advances, if any, and secondary advances in knowledge as well as specialized information in case of particular subject fields. For this purpose, the document has to be read completely, especially where the author states the purpose, method, results, and conclusions of his study.

A word of caution here -- the abstractor has to identify the achievement rather than what the author tried to achieve but did not accomplish or which he proposes to achieve in the future. In addition, the abstractor should identify the audience for whom the document is meant.

It is essential that the document be read completely in order to identify the focus. It is incorrect procedure either to depend on section headings or to start writing the abstract while reading the document concerned.

Perhaps, this step may not be applicable to an author writing an abstract of his own document.

3.5.1.3. Content Analysis

Here, the abstractor analyses the thought / information content recognizing the relationships between the concepts discussed and notes them down without omitting any of them. The language used should be precise (unambiguous) and concise, i.e., free from any verbosity.

3.5.1.4. Assessment

The third step involves the work of assessment of information collected (in the previous step) for its usefulness. This is an important step as it would determine what should be included in the abstract. It is implicit that the information selected should be of benefit to the user for whom it is meant. One of the purposes of assessment is to achieve a certain degree of brevity in the abstract.

3.5.1.5. Synthesis

In this step, the abstract is prepared by synthesizing in a cogent and coherent manner the information collected as a result of the previous steps. For this process of synthesizing, it will be helpful to provide a set of criteria / rules. But, the rules formulated should be applicable to all types of situations, whether it is the preparation of indicative or informative or critical abstracts. This would include one set of rules applicable to all types of abstracts and another set of special rules to meet with the different requirements of the different types of abstracts. While the general rules pertain to structure, language, style, etc, the special rules pertain mainly to the Body section of an abstract. (A detailed description of the rules can be found in *S. Seetharama: Information consolidation and repackaging, 1997.)*

3.5.1.6. Checking

On the completion of writing the abstract, the abstractor should check again for any errors, omissions of information, deviations from prescribed policies and rules, etc. This is a very important step in that if the reader is surprised by what he finds in the original, he will have no faith in the abstracting personnel / service.

3.5.1.7. Editing

In spite of the use of various rules and criteria, many of the abstracts will not be suitable for publication. It is said that less than 1% of the abstracts prepared are suitable for publication without editing. The reasons may be either deviation from the criteria or omissions and in some cases errors. While elimination of errors is vital, there is more to editing than correction of mistakes. Editing seeks to establish cross-references when necessary.

Editing consists of two operations checking and editing proper.

3.5.1.7.1. Checking

The function of checking can be done by non-technical personnel and does not require subject knowledge. In this process, the person should compare the abstract with the original document looking for information that can be checked: document, author and his affiliation, citation, etc. It should be remembered, however, that the checker should neither add nor modify the meaning of the abstract.

3.5.1.7.2. Editing proper

Editing proper consists of the function of elimination of errors apart from maintaining consistency and uniformity in style and presentation. Errors may be errors of format and / or errors of content.

3.5.1.7.2.1. Errors of Format

For eliminating errors of format, the editor should review the body of the abstract to ensure proper grammar and diction. For this, the following operations need to be done:

1. Correct all errors;
2. Singular subject with plural verbs, or the reverse need to be changed;
3. Split infinitives and dangling expressions;
4. Check spelling of words; and
5. Check for correct punctuation.

3.5.1.7.2.2. Errors of Content

Invisible errors of content pose problems. These are of two types - -omission of abstractable material, and nomenclature. To avoid these errors, the editor should:

- Develop the ability to find clues that indicate missing data;

- Ensure the proper and consistent use of nomenclatures in the abstracts.

In addition to the above, the editor of an abstracting periodical should categorise and cross-reference the abstracts and prepare the necessary indexes.

3.6. Appraisal of Information Collected

At this stage, the author responsible for the TC scans through the information collected for relevance or otherwise. In this appraisal, the author will do well to keep in mind on what the users' value in the TC.

A set of criteria used by users in the evaluation process are given below:

- **Information / data quality :** precision, accuracy, credibility of source(s), recency.
- **Scope :** completeness, comprehensiveness of subject coverage.
- **Appropriateness.**
- **Haggle and hassle factors** like time lag, ease of usage of information, ease of access.
- **Costs :** Direct cost / Indirect cost in accessing and using the service and in post-processing of information.

While in case of ambiguity, the author may ask for the original documents, in the case of documents in languages with which he is not familiar he would request for translations. In an attempt to satisfy the author's requirements, the library personnel would interact with the translators and reprographic specialists.

3.7. Scope Redefinition

On the basis of information collected, the author in consultation with the subject specialists / peer groups and readers (user groups) may redefine the subject scope of the TC -- either to enlarge or reduce the subject scope -- if necessary. In this step, all the specialist groups who were involved in scope definition earlier would be consulted for scope redefinition.

3.8. Draft Text Preparation of TC

Draft text preparation involves the arrangement of ideas (working units of information) embodied in the abstracts collected in a helpful sequence and integrating the information collected into an organized text. For this purpose, it is necessary to prepare a Document Plan before attempting to write the first draft.

3.8.1. Document Plan

The Document Plan consists of:

1. Defining Requirements : Subject overview; Scope and purpose of the document (why is it needed? What is to be accomplished? How will it be used? What will it not do? How does it relate to other and information needs of the intended audience)?

2. Outline Preparation : Detailed listing of the major sections and subsections. A Visual Table of Contents (VTOC) may support the topic outline (In developing a topic outline, it may be helpful to use a model and also select an organizing principle).

***Note* :** In the development of a Topical outline, the principle to be adopted must be compatible with the document objectives. Generally speaking, four main organizing principles can be used. They are:

- Logical organization.
- Chronological Organization -- is time ordered.
- Psychological Organization -- is results oriented. It covers the most important topics first. This organization provides a stronger presentation for a long-range strategy report, for instance, because it presents recommendations and conclusions first. Marketing materials stress benefits, and a psychological organization states these clearly.

- **Topical Organization :** here, the order of sections is not very important as long as all information on a topic is together. In this organization, one selects topics that logically group information. A technical newsletter with regular columns on new developments, Operator tips, User support, and Personality or Location Profiles is arranged topically.
- **Specifying the Schedules and Resources :** support service requirements (Word processing, Reviewers, Case study requirements, Development schedule, Number of copies required, Distribution requirements). (*Sandra Pakin & Associate. 1984*)

This Document Plan needs to be reviewed and approved before writing the first draft.

3.8.2. Arrangement of ideas

One and the same set of ideas can be arranged in several ways. The choice of a particular sequence should be based on its helpfulness to the user. This depends on the specialization of the reader.

For example, a trend report on "Metabolic Disorders" can be organized in the following two ways:

Table C2: Arrangement of Ideas

By Substance affecting	By Organ affected
Protein disorders	Whole body
• Hyperproteinemia	• Hyperproteinemia
• Hypoprotenemia	• Hypoprotenemia
• Myeloma	• Obesity
• Haemoglobin disorder	
• Amino academia	Cell
• Phenylketonuria	• Protein deficiency
• Alkaptonuria	• Nucleic acid disorders
	• Lipid storage diseases
	• Glycogen storage disease
Nucleic acid disorder	Tissue
• Gout	• Alkaptonuria

- Lupus erythematosus

Lipid metabolic disorder

- Obesity
- Hyperlipemia
- Lipid tumours
- Lipid storage diseases

Carbohydrate metabolic disorder

- Glycogen storage disease
- Galactosemia

Joint

- Gout

Heart

- Arteriosclerosis
- Myocardial infarct

Blood

- Hemoglobin disorders

Liver

- Protein deficiency
- Liver cirrhosis

Pancreas

- Diabetes mellitus
- Protein deficiency
- Phenylketonuria
- Lipoid nephrosis

Spleen

- Hurler-Pfaunda disease

Skin

- Lipid tumour

Child disease

- Galactosemia

The choice between the methods of grouping of the ideas mentioned above in the form of informative abstracts depends largely on the orientation required to meet the needs of the reader. If the person is interested on "substance causing metabolic disorders", the headings and arrangement given in column 1 will be more helpful. On the other hand, if he is interested in the study of "organ affected" arrangement given in column 2 will be more helpful. Thus, the grouping of ideas into chapters depends on the primary focus of a subject in which the reader may be interested.

The ideas grouped under each of these headings are themselves to be arranged in a sequence helpful to the majority of readers of the concerned TC. Such an arrangement of ideas starts by arriving at the sequence of group headings which ultimately become either chapter headings or part headings. The process can be aided by using any one of the principles mentioned in Sec 6 of Chapter B. For example, in the Table, the sequence of ideas mentioned in column 1 are arranged on the basis of the principle of increasing complexity of the disorders caused by the different substances. If preferred, the symmetrically opposite sequence can be secured.

At this juncture, it may be mentioned that the choice between different methods of grouping of ideas depends largely on the orientation required. It is here that the experience gained in the earlier step Scope determination and the later step Scope redefinition comes in handy in the arrangement of ideas in a particular sequence.

3.8.3. Integration of Information

The information collected from various sources, after analysis and arrangement of ideas in a sequence helpful to the readers /users and with appropriate orientation are to be integrated or woven into an organized text, if the TC is a state-of-the-report, review, technical report, etc. For this purpose, the target audience and their intellectual level or standard, the intended use, the impact one desires on the audience, the peculiarities of the subject, readers' preference to a particular kind of presentation are some of the facts that need to be considered. The result of integration should conform to the Principle of Unity of Ideas. In this process of integration, the objective of the TC should be kept in mind. The flow of ideas should be logical, step by step through the structural elements of the text. In addition, the presentation of information should be based on an optimal mix of text, illustrations, etc. This is especially so for planners, decision-makers who are hard up of time and hence need information presented in condensed form like tables, graphs, charts, diagrams, etc. On the other hand, in the case of extension workers, information should be presented in an easy-to-understand form and should be well illustrated.

In relation to dynamics of consolidation, Gopinath (1981) has given a typology for the consolidation / integration process which can be used in the organization of texts. He has identified ten types of organization of texts:

1. Trends or occurrences in a series of happenings over a period of time, resulting in a chronological arrangement of events.
2. Trends or occurrences on either side of a watershed event.
3. Trends against a background of a developing situation or crisis.
4. Cycle of phases in a maturation process.
5. Continuum between two extremes.
6. Two axes, enabling two dimensional plotting between sets of extremes.
7. Ranking order, series arranged between high series.
8. Yardstick /criterion (Ranking order used as basis of comparison for a series.)
9. Geo-historical survey or the like (comparison of events in a range of countries over a period of time).
10. Profiling (Rating composite set of scores on a number of inter-related dimensions).

Saracevic & Wood suggest that restructuring of information should be in consonance with the communication practices of readers / users on the one hand, and on the other should reliably reflect the state of knowledge or data on the given subject. Elaborating, they observe that more specifically, restructuring requires determination of:

- Extent to which available content is to be incorporated;

- Degree of invariability of information as found in the original (i.e. which new information value is to be added);
- Degree of detail in information;
- Degree of changes in sequence of presentation relative to the original;
- Technical sophistication; and
- Editorial qualities.

To summarise, the integration of information can be achieved by:

- Application of Classificatory Principles.
- Adopting different methods of presentation of information.
- Application of Visual Communication Techniques.
- Following general rules of grammar, style, etc involved in technical writing.

3.8.3.1. Application of Classificatory Principles

In Sec 3.8.2, it has been shown how arrangement of ideas is facilitated by the application of classificatory principles. The resultant arrangement can then become a framework round which information collected can be integrated into an organized text. The framework can be used for packaging and repackaging of information into an organized text. In other words, an analytico-synthetic approach is adopted for the purpose of integration of information.

Advocating a similar procedure, Saracevic & Wood (1981) have suggested the following steps for the purpose of integration:

1. Sorting of extracted information into classes and subclasses (headings and subheadings) according to the table of contents, classification scheme or typology for the subject or mission;

2. Comparative arrangement and merging of extracted information within each class and subclass;

3. Comparative evaluation of different extracts or data in each class and subclass;

4. Resolution of conflict (if any), or decision to present conflicting information

5. Compression of information into a structure and form most suited to intended users and uses and in accordance to objectives, resources, constraints of the system or unit as a whole.

6. Evaluation of the final product according to criteria related to users and use.

In 1980's, Ranganathan's Generalized Facet Formula based on Postulates and Principles as enunciated in the General Theory of Classification has been applied in the preparation of Technical Information Packages on Biogas Technology, Cook Stove Technology and Wind Pumps. Two Manuals on Biogas technology for Managers and Extension workers have been prepared separately, extracting the relevant information from the Biogas Technology handbook prepared earlier.

3.8.3.2. Adopting Different Methods / Modes of Presentation of Information

Broadly speaking, two modes of presentation of information are possible. They are written and oral presentations. As TCs generally adopt written mode, only the various written modes of presentation are discussed here.

Written communications are preferred as it ensures better control on the one hand, and also as it allows modification and correction wherever and whenever it is necessary. Basically, two types of written communications / presentations are encountered — namely, Textual (words) and Non-textual.

3.8.3.2.1. Textual Presentation

In the case of a TC, the presentation of information may differ according to the target audience concerned. Thus, the presentation of ideas and lay-out of the text may have to be altered depending upon the destination. For example, in a review primarily meant for Top Management, the conclusions and recommendations have to be highlighted and placed as the First Section of the report, followed by a detailed data and / or amplifying information, i.e., a *Finding-Oriented Arrangement.* On the other hand, in describing experimental work, the sequence should be Purpose, Method, Results, Conclusions, Collateral and other information, i.e., *Experiment-Oriented Arrangement.* However, the best possible sequence is dependent on the readers to whom the TC is meant.

Therefore, the author should avoid the natural tendency as a writer to think of primarily himself and thus write primarily for himself. He should use consciously language that not only facilitates communication of ideas, but should create the desired effects in the user to act. For this purpose, the author should cultivate a psychological sense, that is, sensitize himself to the readers' needs and responses.

Some questions that the author can ask himself (imagining himself to be the reader) are:

- Is his style too complex to be readable or too plain or is it just right and why?
- What is his tone, and how does he achieve it?
- Do I like it or don't?

In other words, the author should be conscious of the message to be conveyed when using language as the medium for communication. Some of the points that can be considered are:

a) Use language that sells -- vigorous verbs, strong nouns, assertive phrasing, etc.

b) Phrasing thoughts clearly.

c) Anticipating the many questions, and responses of the readers.

d) Offering him variety to keep him interested.

This essentially means that the language used should be oriented to the intellectual level, comprehension and requirements of the users concerned.

3.8.3.2.2. Non-Textual Presentation

Non-textual presentation can be in one of two forms, namely, Tables and Illustrations / Figures. Tables are distinguished from illustrations in that the tables contain only data arranged in systematic order within columns and rows; whereas illustrations include graphic materials like charts, graphs, diagrams, plates, etc.

Tables : Tables are tools of analysis and communication that facilitate understanding of the fullest meanings of complex patterns or relationships. However, when information in a table can be further processed to result in some conclusions / inferences, an annotation should accompany the table concerned.

Illustrations : Illustrations are also referred to as figures and occur in technical books & periodicals. Unlike tables, illustrations are difficult to make and need professional expertise & money.

An illustration or figure is defined as any kind of graphic representation which cannot be typed or printed (Cambell & Ballou, 1978.). It should present only one fact or series of related facts and should be easy to interpret.

Each type of illustration is useful in its own way. For example:

Orthographic drawings are useful for engineering description and record; for conveying all dimensions, specifications and other necessary data as a guide to machinists and other production workers for fabrication.

Schematic diagrams, on the other hand, are helpful in illustrating function, path of flow, essential lines of interconnection among components of a system and terminal points. They can also be used to depict electrical and electronic wiring, architectural piping; transportation routes, system flow; industrial production sequence and flow, etc.

Charts & Graphs are useful as aids in order to:

- Indicate significant trends of an activity over a period of time.
- Provide better management visibility of finances, production, quality control, marketing and the like on a regular basis.
- Plan & schedule programmes & products through the use of PERT, graphical systems.
- Supplement verbal discussion in a meeting.
- Air programmes of technical training, new employee orientation and the like.

Photographs are useful:

- To authenticate and certify the information contained in the text.
- To make the presentation effective.
- To face a deadline.
- As filling material in a technical presentation.

However, one of the problems encountered is the decision regarding when one should resort to textual presentation in preference to non-textual presentation. If the latter is preferred, then the question arises as to whether it should be tabular or illustrative.

The situations when one can resort to either textual or non-textual presentation are indicated in the table below:

Table C3: Written Modes of presentation

Textual presentation	Tabular	Illustration
When Textual?	*When Tabular?*	*When Illustrative?*
• When detailed information is descriptive, explanatory, narrative, etc	• When large amount of data is to be presented within a short space & with clarity	• Do words alone tell the story?
• When information cannot be segmented into a pre-determined format.	• When comparison of data/information is required	• is an illustration needed to: * Provide a clear understanding. * Amplify & stress on a central theme or idea expressed in a text. * Record & document, a method, design or event?
• When the subject is predominantly verbose and non-numeric	• When categorization of data/information is required.	
• Steps of logic and presentation of abstract ideas.	• When evaluation of data/information is required.	

3.8.3.3. Special Orientation

In the case of a technical review, the presentation of the text may differ according to the target reader / user to whom it is primarily meant. The destination of the document may be a particular enterprise, such as a university, government, or any department of either. A review may have different destinations even within an enterprise -- it may be for any one of its Divisions-in-charge of Planning, Design, Production, Stores, Finance, Accounts, Sales, Public -- may have to be altered depending upon the destination. For example, a review meant for Top Management, the conclusions of the investigations may nave to be highlighted and placed as the very first section and the detailed data may have to be appended to it.

In the preparation of the draft text, the author plays the dominant role in the arrangement of ideas and integration of ideas into an organized text. Others, such as, Library & Information Science personnel, Communication and Information Presentation specialists, and readers / end users / intermediaries play their respective supporting roles in the draft text preparation.

3.9. Testing of Draft Text

Evaluating the quality & effectiveness of a TC is an essential step in its preparation. In order to do so, the draft text is:

1. Reviewed and assessed by subject experts; and
2. Field-tested by using it with the ultimate end users / beneficiaries or intermediaries. However, before testing the draft text prepared, it would be a wise policy to keep the manuscript in 'cold storage' for sometime and then to revise, if necessary.

3.9.1. Review and Assessment

The TCs may be sent to subject experts in the respective areas. The factors to be considered in the review & assessment are:

- Technical Accuracy
- Completeness
- Perspective
- Analysis
- Synthesis
- Reasonableness of organization
- Appropriateness
- Utility
- Value-added

It has been suggested that in reviewing, all aspects of draft need to be reviewed and not just spelling, punctuation and grammar. As a matter of fact, one has to look for a complete range of problem areas like:

- Interest level, adaptation to audience
- Persuasiveness, purpose
- Content, organization
- Clarity of discussion
- Coherence, use transition
- Title, introduction, and conclusion
- Sentence style and clarity
- Handling of graphics, etc. (McMurrey David A).

As a matter of fact, the TCs on Biogas Technology, Cook stoves, and Wind were sent to experts:

- To examine the comprehensiveness of the TCs;
- To verify the factual information; and
- To know what they feel about the TCs.

3.9.2. Field Testing

Field testing the draft text facilitates improvement in the orientation of the text as well as ensures user acceptance and effectiveness of the document in the operating environment. Some of the questions that can be asked of the participants are:

a) What is your general opinion about the TC?

b) For what purpose was the information in the TC useful?

c) What specific information was of special interest?

d) What sections were difficult to understand?

e) Was it easy to find information?

f) What are the errors?

g) What terms / topics need explaining?

h) What procedures were hard to follow?

i) Was any information lacking / missing? If yes, what additional information should be included in the TC?

j) What did you like best about the TC? What least?

Note : *It would be useful to distribute a proforma for communicating the feedback. Possible topics for comment should include: clarity, questions not answered, accuracy, organisation, exhaustiveness, figures, examples, format, usefulness of text, etc.*

3.10. Finalisation

Based on the feedback received in the previous step, the author should finalise the text of the TC incorporating the changes suggested. These changes may be -- addition or omission of information; changes in the sequence of ideas presented; change in the form of presentation -- i.e., use of diagrams, graphs, tables, etc instead of descriptive paragraphs, etc.

3.11. Organising of the Technical Communication

3.11.1. Style Guide / Manual

A Style Guide or Style Manual is a set of standards for design and writing of documents, either for general use or for a specific publication or organization (Wikipedia). It is a document that prescribes the 'styles' or standard format for the presentation of information & deals systematically with the presentation of content in a particular publication by identifying standards that the publisher wants to apply (Laidlaw). It has also been described as a reference document that includes rules and suggestions for writing style and document presentation and often specifies which option to use when several options exist. In addition, they include items that are specific to the Organization concerned and items for which a "standard" or example does not exist. However, the specific content of the style guide is not usually a matter of "correct" or "incorrect" grammar or style, but rather the decisions one has made from among the many possibilities (Weber, 2011). It essentially provides writing rules for a particular academic discipline or organization and includes information such as, how to present the text, tables or illustrations; how to cite research within the text of the TC; what to document, how to abbreviate, how to write headings within the paper, etc. Some style manuals also include sections about how to approach the subject matter of your TC, how to select a topic, use the library, compile a bibliography, and take notes, etc. (Oregon State University).

In other words, when one uses a style manual, the writing style will be consistent throughout the TC and will be understandable to the reader, since he also will be a scholar who knows these standards.

3.11.1.1. Purpose / Role of a Style Guide

According to Laidlaw, the key purpose of a style guide is to make life easier by reducing the amount of thought one has to put into tiny details on an ongoing basis. It may be said that a style guide:

- Makes content preparation easier.
- Makes content maintenance.
- Supports design & marketing objectives.
- Helps portray a professional approach.
- Makes content more usable.
- Makes it transferable across all publications

On the other hand, Weber has identified several purposes that style guides can serve. They are:

- To ensure that documents conform to organizational image and policy, including legal requirements
- To inform new writers and editors of existing style and presentation decisions and solutions

- To define which styles are negotiable and which are not
- To improve consistency within and among documents especially when more than one writer is involved or when a document will be translated
- To remove the necessity to reinvent the wheel for every new project
- To remind the writer of style decisions for each project, when one writer works on several projects that have different style requirements
- To serve as part of the specifications for the deliverables. When writing for clientele outside the organization or when outsourcing writing projects.

Other purposes of style guides stated are:

- Serve as training tools
- Provide documentation of the writing & publishing process
- Can be used as a marketing tool (Elridge).
- Can present publications in a consistent and readable form
- Helps to locate rules about how to present the form and shape of your paper, thesis, etc. (Oregon State University).

In short, the benefits of a Style guide are:

- To the Organization:
  - — Consistency in style and format
  - — Professional look for documents
- To users:
  - — Answer common questions
  - — Training & skill development
- To the Customers:
  - — Quality assurance / Value-added (Elridge).

As per Wikipedia, Style Guides are useful for:

- Academia & Publishing : documentation, formatting style for citations & bibliographies, for submitting articles and communications to academic journals, etc
- Preparing Newsletters, News releases, Websites
- Demonstrating the preferred lay-out and formatting of a published page in Newspapers, Graphic Design.
- Legal Documents
- Journalism
- Technical Writing
- Electronic Publishing.

3.11.1.2. Style Guides / Manuals Available For Academics

The following are some of the Style Manuals available:

a) A Manual for writers of term papers, theses & dissertations *by* Kate L. Turbian.

b) ACS Style Guide. (American Chemical Society).

c) American Medical Association Manual of Style.

d) Geosciences Reporting Guidelines.

e) IEEE Style.

f) MHRA Style Guide.

g) MLA Style Manual. Edn 3 and MLA Handbook for writers of research papers, Edn 7.

h) Scientific Style and Format: The CSE Manual for authors, editors, and Publishers. Edn 7.

i) The Chicago Manual of Style.

j) The Style Manual for Political Science.

3.11.2. How to Write a Scientific / Technical Communication

One of the initial questions encountered by a writer is "How to start and what should be the format". Other questions may be "what content to include and what not to include?" Perhaps, a Scientific Format would be ideal as it would effectively communicate the findings to all the readers in a uniform manner. Additionally, it would facilitate the reading of the TC at several different levels -- be it the title, abstract, results, or tables and figures, etc. In other words, the Scientific Format ensures that whatever levels a person reads, he is likely to get the key results and conclusions. The Format , therefore, should include the following sections: Title, Author(s) and affiliation, Abstract, Introduction, Methods or methodology, Discussion, Acknowledgements, Bibliographical References / Literature cited, Appendices (if any), Tables, Figures, etc. (AIP Style Manual).

3.11.2.1. Pre-requisites for Writing

To ensure that the contents of the TC are logical, readable and understandable, the following steps or pre-requisites become necessary:

1. Define the requirements:
   a) What type of TC are you writing? Subject Scope and Purpose of the TC?
   b) *Audience definition* -- background, experience, level, expectations, knowledge of the subject discussed in the TC, and their information needs.
2. **Outline preparation :** Logical sequence in which information should be presented; Detailed listing of Chapters, Sections, Sub-sections i.e. a sort of Visual Table of Contents (VTOC) indicating the likely Divisions /Chapters, Sub-divisions / Sections and sub-sections would facilitate writing the TC; & What information should be presented in each chapter / section of the TC?

   (Obviously, this VTOC may have to be revised time and again while writing.)

3. **Plan for Textual and Non-textual forms** — tables, graphs, figures, illustrations, photographs, etc, if any — of presentation of information.

3.11.2.2. Actual Writing of the TC

3.11.2.2.1. General Considerations

While preparing a TC, it is essential that initially one decides on the type of TC and what to cover & what not to cover in relation to the topic or to be modified as and when necessary while actually writing. Obviously, a plan needs to be made wherein key issues are considered. The next point for consideration is the order of writing --- first chapter first followed by others in a sequence 2, 3, 4, etc or the most difficult or easiest first. Perhaps, it would be a wise policy to write the Introductory and Conclusion chapters last as by then the writer would know the content covered of the TC being prepared.

On the question of content, the text should have covered the background of the study, the methodology / methods adopted for the study, and what were the inferences / conclusions and recommendations made. This approach will facilitate chapterisation -- for example, Chapters 1 & 2 may cover the background of the study, Chapter 3 the methods of the study, and Chapters 4 & 5 the inferences / conclusions drawn and recommendations suggested. In this context, it is worthwhile mentioning that one can prepare a worksheet collating / mentioning information required for different Chapters as it would facilitate consolidation of information keeping in mind the Principle of Unity of Ideas. However, it should be remembered that for effective communication, the language (style, grammar) used should be precise, concise and clear. In addition, in conformity with the concept of Technical Writing, the text should be oriented to an educated audience, not necessarily intellectuals like academicians and /or scientists. In other words, the effort should be for maximum usability of the content by a large number of readers / users. One word of caution - - - write what you mean and not what you don't mean i.e. it should be free from any ambiguity whatsoever. Further, as a rule, the theme should be reflected in the TC from the beginning to the end. In order to achieve this, the entire presentation should be in a logical sequence with connectedness all through, though the information presented is compartmentalized into chapters / sections and sub-sections.

Another factor that needs to be considered is Typing Specifications like Type Size, Font. Chapter Head Pages and Page Layout. Margins. Format of Thesis, Facing Pages. Use of Single Spacing. Page Numbers. Page Number Location, etc. This is to ensure consistency and adherence to typing specifications.

(*Details on typing specifications are not given here. The reader / writer may consult a* standard *style manual / guide for this purpose*).

Morrison (2004), in his article on "Tips on scientific writing", has given the following simple principles:

1. Don't "write". Rewrite - - Plan from the outset to revise your paper many times.
2. Do one thing at a time as writing a paper requires lot of steps for which one needs to organize one's ideas.
3. Develop your paper from the inside out - - - start with the Body of the paper followed by conclusions with introduction being written last.
4. Don't try to write "perfect" early drafts, but those which can be polished later into final drafts that can be readable and free of grammatical errors.

3.11.2.2.2. Format/ Structure of TC

According to AIP Style Manual), the Format should include the following sections: Title, Author(s) and affiliation, Abstract, Introduction, Methods or methodology, Discussion, Acknowledgements, Bibliographical References / Literature cited, Appendices (if any), Tables, Figures, etc.

On the other hand, Kim Kastens and others have suggested the following:

- Title Page
- Abstract
- Table of Contents
- List of Figures
- List of Tables
- Introduction
- Methods
- Results
- Discussion
- Conclusions
- Recommendations
- Acknowledgments
- References
- Appendices

Depending on the author, institution, and authorities concerned, the appropriate format can be designed and followed.

3.11.2.2.3. Steps in Writing

Based on the methodology described in this chapter, one can proceed with the actual writing of the TC. It may be mentioned here that, generally, a TC consists of three matters, namely, Front matter, Text or Body, and Back matter. The following are the steps in Writing:

FRONT MATTER

The front matter includes items such as Title page, Abstract, Table of Contents, Lists of Figures, Illustrations, Abbreviations, Tables (in no particular order), Preface or

Acknowledgments (optional).

1. Prepare the Title Page : The Title Page should contain the following elements:

Title of Document : The title chosen should coextensively express the thought content of the document being prepared. It may have a Main title followed by an explanatory title, if necessary. Fanciful & misleading titles need to be avoided; ***Author(s) and affiliation; Publisher & Place of Publication (if any); Year of Publication.***

2. Prepare an Abstract : Depending on the document concerned, an informative or Indicative abstract has to be prepared. An Informative abstract of about 250 to 300 words needs to be prepared following ***The Guidelines for Abstracting*** given in Chapter B, Sec 3.5.The informative abstract should contain sufficient details on purpose, methods, results, conclusions and recommendations (if any), so that reference to the original will be minimal. On the other hand, indicative abstracts would be useful in relation to review documents and must indicate the contents only.

3. Table of Contents : It is based on the topical outline of the TC developed earlier and is essentially a compilation or list of headings and sub-headings with page numbers. It would be helpful & useful if the sub-headings are indented. One should be consistent in the level of heading that one lists; e.g., if you list the second-level subheadings from one chapter, you should list the second-level subheadings from all chapters that contain this level.

4. List of Figures & List of Tables : Use a separate page for each list. List the number, caption / short title, and page number of every figure and table in the body of the TC.

5. Acknowledgement(s) : A separate page for

expressing professional and personal indebtedness may be included.

TEXT OR BODY

The text or body is usually divided into chapters -- *The number of Chapters depends on the subject and author concerned. However, the content of the Chapters, as a whole would cover items 6 to 11 mentioned below* -- and provided with introductory and concluding sections, which may or may not be designated as chapters. In each chapter, subheadings indicate the orderly presentation of the concerned information. For this purpose, an appropriate system of headings should be formulated and used throughout the TC.

6. Introduction : Begin with a provocative statement that will stimulate the interest of the reader to go through the rest of the TC. It may concern the problem or issue that you are focused on in the study. It should motivate & draw the attention of the reader and also stimulate him to read further. Obviously, the introduction should provide the background to the study as also the work done earlier in regard to the problem being studied. For this purpose, reference to libraries and reference sources like encyclopedias, handbooks, etc may be useful & helpful. In addition, a definitional analysis highlighting the definitions of the terms used & their implications needs to be given. Technical glossaries would be helpful in this regard. The following sections, depending on the choice by the research scholar, need to be mentioned **briefly**:

- ***Proposed Research / Study*** **:** highlighting the need for research / study on the chosen topic / subject; the statement of the problem; the specific objectives in relation to the problem, including any prelude to the research investigation, if any;
- ***Methodology / Procedure*** **:** to be adopted in the research investigation is mentioned briefly, with

minimal details like the choice of the specific topic, the procedural steps to be followed, hypotheses, if any, limitations of the study, the analysis of results / observations, and a summary of findings.

- ***A Conspectus*** **:** indicating briefly the contents of the rest of the chapters / sections.

The Chapters may cover:

a) A review of literature on the topic chosen;

b) Systematic study of the subject presenting details of the scope of the subject as a whole, main divisions and sub-divisions (if necessary), terminology, landmarks in the evolution of the subject, state-of-the-art and trends, etc ;

c) A detailed account of the study undertaken and observations & results seen;

d) Discussion followed by Conclusions and Recommendations, implications, if any)

7. Methods : Include description of Materials, Procedure, Techniques, Equipments, Theoretical / Experimental/ Analytical methods, Assumptions, Limitations, etc. Provide all information necessary that would facilitate repeating and / or evaluating the study.

8. Results : Write / Include actual statements of observations including statistical; Tables & Graphs. Both positive and negative results, if any, need to be included. One word of caution -- do not interpret results. If amenable, present results logically using sub-headings. Plan the results section to achieve the most effective mix of text, figures, and tables in the presentation of the findings.

9. Discussion : Begin the discussion with a brief summarization of the important /significant results. The discussion should centre on the major patterns of observations; relationships, trends and generalizations; likely causes / mechanisms for the results observed;

interpretations & implications of the results; correlation with earlier results; and significance of the results.

10. Conclusions : Based on the results and discussion, draw inferences/ conclusions explicitly in a simple language.

11. Recommendations : Suggest action(s) or follow-up research studies to be done as well as indicate future work to be done on related topics.

BACK MATTER

The back matter consists of Bibliography or reference list (may be in back matter or at end of each chapter), Appendices (if any, they may come either before or after references), Endnotes or Notes (if any, they may be in back matter or at end of each chapter).

12. Bibliography or Reference List : Include all documents referred to in the text, as also other documents consulted while preparing the text of the TC. Each reference to consist of a serial number, chapter/section of the text where referred, name author(s), title of the document, name of journal /periodical, volume number, issue number, month/year of publication, inclusive pagination. Alternatively, follow the style suggested by the journal editorial committee or any authoritative standard of a professional body like Bureau of Standards. Completeness, accuracy, and consistency of references are more important than strict adherence to a particular citation system. Using the information provided in your list of references, the interested reader should be able to locate each document that you cite in your paper.

13. Appendices : Include all material related / relevant to the Text like questionnaires, detailed raw data and / or material too long for inclusion in the text, lists of institutions, experts, byproducts, etc. The appendices should be uniquely designated as A, B, C, etc and should have a title.

It may be pertinent to mention here that for achieving an effective style in writing, the points outlined in section 3.8.3.2.1 may be adhered to. Alternatively, the writer may choose to follow any Style guide or manual. Perhaps, what is important is that the TC prepared must be readable, understandable and usable.

4. MANPOWER FOR TECHNICAL WRITING

In Chapter B, Pre-requisites for TW, the knowledge and skills, in general, required for TW have been discussed. However, the knowledge and skills required by librarians and information professionals at each step in the technical writing process is indicated below:

Table C4: Knowledge and Skill for Technical Writing.

S. N.	Step / Stage	Knowledge of	Skill/ Ability
1.	Recognition of the need for a TC:	User studies, Reference Framework, Data / Information Analysis,	To interact with users & recognise their needs
		Highways & byways of subject(s).	
	& Audience analysis:	Attributes of users. & methods of study	To specify target group
2.	Choice of Subject of TC/ conceptual analysis of the Subject:	Study of subject method, Subject/ information Analysis	To specify subject precisely to be covered by TC.
3.	Team formation for design & development of TC:	Preparation of Directories of institutions/experts.	To identify & organize a team of specialists
4.	Collection of Information: (pertaining to the subject(s) to be covered by the TC).	Core documents	To identify sources of information.
		Abstracting & condensation techniques	To identify & select relevant information
		Classification & Indexing	To technically process information

5. Appraisal of information Collected:	Criteria for assessing information. Evaluatory techniques.	To evaluate information
6. Scope redefinition, if necessary.	As in step 2.	To redefine the scope in consultation with experts
7. Draft Text preparation	Consolidation and Integrating techniques. Guiding Principles for: Idea, Verbal & Notational Planes. Modes of presentation. Bibliographical Standards. Supplemental aids & their preparation.	To consolidate & integrate the information into an organized text.
8. Testing of Draft Text	As in Stage 5. In addition, Feedback techniques.	To evaluate the technical writing
9. Finalisation	Standards to be followed	To incorporate the changes suggested. To prepare the supplemental aids necessary for the users.

Chapter D
Organisation And Presentation of Information

1. ORGANISATION OF DATA / INFORMATION

Raw data collected during scientific investigations which need to be transformed into some format that allows interpretation and analysis between the variables. In other words, any data collected in research or reference is always in an unorganized form and needs to be organized and presented in a meaningful and readily comprehensible form in order to facilitate further analysis and interpretation. Data collected can be presented in the following ways:

Classification — is the process of arranging entities according to their resemblances, similarity or identity. It can be *chronological or temporal classification; geographical or spatial classification; qualitative classification based on some attributes like sex, literacy, religion, etc; quantitative classification based on characteristics which can be measured and numerically described such as, height, weight, age, income, etc.*

Tabulation — is the process of summarizing

classified group data in the form of a table. *A table is a systematic arrangement of classified data in columns and rows.*

Diagrammatic Presentation — is one of the most convincing and appealing ways in which data may be presented is through charts. *A chart can take Circular & Pie diagrams, Time series or Line graphs, Frequency graphs.*

(Vide Section 3.8.3.2.2. of Chapter C.)

2. PRESENTATION OF DATA / INFORMATION

2.1. Parameters

One of the main reasons for handling data or information at work is usually an objective which involves influencing or affecting other people in a desired way (Parker, 1983).

Information may be meant for:

- **Stimulation** : to help one to be creative and motivated or interested
- **Knowledge** : to get competence
- **Current Awareness** : to keep one up-to-date with the competition and developments
- **Productivity** : to tell one how to do something efficiently and effectively
- **Theorising** : to predict different or future phenomena
- **Comparison** : to compare one's own results or use in consultation
- **Problem-solving** : to show how other people have tackled similar problems
- **Education** : to give background details in a new field or to pass an examination
- **Decision-making** : to be better informed about alternatives and consequences

- **Adaptation :** to adjust to demands of and changes in the environment

At the same time, it is emphasized that communication which is essential to our way of life is dependent on the method adopted for communicating one's own work. In other words, one has to decide whether it would be written or verbal communication. In this context, one should find answers to the following questions for taking a decision:

1. How do you wish to influence them?
2. What level do they understand?
3. What will they use the information for?
4. How much time will they spare to read it?
5. How do they prefer to receive information -- a long essay, drawing, brief, clear notes, a talk, etc?

Another aspect that needs to be considered in the Presentation of Information is the subject of the document. Depending on the nature of information, an appropriate system of presentation should be chosen. In other words, different presentation techniques / forms are adopted within one and the same subject. In relation to each of the forms, certain conditions should be satisfied before they can be adopted.

One other aspect which influences the form of presentation relates to the type of TC being prepared. For example, for a state-of-the-art-report, textual presentation should be adopted. However, if necessary, other forms may be used to supplement the textual presentation. On the other hand, manuals, guidebooks are better illustrated with minimal textual presentation of information rather than full textual presentation.

The type of presentation is also influenced by the type of users to whom it is meant. For a manager hard up of time would prefer graphical, tabular, or pictorial presentation with minimal textual presentation. On the other hand, R & D scientists would prefer textual presentation with necessary illustrations. The extension workers who perform with the aid of manuals, guides, etc are comfortable with illustrative manuals than fully text-oriented manuals.

From the foregoing, it becomes clear that the parameters to be considered in the design and preparation of TCs are the following:

- User
- Purpose / Impact wished for on the user
- Subject
- Product / TC
- Media of communication — Written or Verbal

2.2. Modes of Presentation

Broadly, two modes of presentation, written and verbal or oral presentation are possible. TCs generally adopt written mode as it ensures better control on the one hand and also as it allows modification and correction wherever and whenever it is necessary. For a full discussion on modes of presentation, the reader may refer Section 3.8.3.2 of Chapter C.

2.3. Presentation & Production of TCs

In the context of TCs, a technical communicator is expected to consider the following points before he makes a presentation:

- Right perspective
- Clinical analysis of the reader
- Reader response
- Statement of objectives
- Selection of single basic thought, reinforced by psychological and physical factors
- Collection of information

- Presentation of ideas
- Evaluation of audience response

It is worthwhile mentioning that an effective presentation is:

- One that
  - — is clearly organized
  - — arrests the attention of the readers
  - — provides appropriate descriptions with illustrations & examples
  - — ultimately gives the readers a memorable and positive experience
- Determined by
  - — writer's unambiguous understanding of readers
  - — abiding interest in readings of writer
  - — quality of contents in terms of authenticity, reliability, authority, adequacy, currency, etc
  - — structural organization of contents
  - — putting organized contents into a language which presents the ideas with clarity & lucidity supported by illustrations and examples.

On the other hand, physical production of resulting final draft in terms of

- Readability
- Print aesthetics
- Positioning of illustrations appropriately to go with the textual matter and other related aspects.

2.3.1. Abstracts

As you are aware, an abstract is the summary of the essential contents of a work, usually an article in a journal, together with specification of the host document and the focus within it. *(For details regarding Guidelines for Abstracting, see Section 3.5.1 of Chapter C).* In the preparation of abstracts, all rules applicable to language, style, and grammar should be followed. Some of the rules in abstracting are:

1. Use simple, precise and concise statements but with some variety in length and structure to avoid monotony. Use complete sentences. Telegraphic language is permissible, but without sacrificing precision and clarity.
2. Use phrases for clauses, words for phrases when possible.
3. Use standardized terminology, abbreviations, acronyms, initialisms, symbols, etc in the abstract.
4. Each abstract should consist of three sections or parts:
   a) Reference Section: Directs the user to the original document.
   b) Body Section: Contains data / information from the original document and / or indicates or describes the content with descriptors.
   c) Signature Section: Identifies the abstractor.

   In regard to the content:

- Include information pertaining to the purpose, method, results and conclusions. Exact boundary of the subject, the primary and secondary advances made, if any, should be indicated. Use of new instrument, technique, aid and outstanding factual data should also be indicated.
- Do not include all data reported in the original, but, only samples and examples of data.
- Avoid old information, details of procedure, and redundant information
- Omit information inferable from the Reference Section.
- As regards arrangement of content, ideas / concepts included in the abstract should be arranged in a

manner to save reader's time. One can adopt **Finding-oriented arrangement** with results and conclusions first followed by supporting details, other findings, methodology, etc i.e. amplifying information, Or **Experimented-oriented arrangement** with the content elements in the sequence Purpose, Method, Results, Conclusions, Collateral and other information. However, the best possible sequence is dependent on the users concerned.

- In case of Informative and Critical abstracts, use active voice and past tense, and for Indicative abstracts, passive voice and present tense.
- In regard to size / length of abstract, 200 to 300 words for informative and critical abstracts, and about 50 words for indicative abstracts.

2.3.2. Technical Reports

A Technical Report is one that presents technical data/information as objectively as possible in a simple, direct, concise and precise manner. The data may be numerical and/or non-numerical. These technical reports may be oral reports, technical articles, proposals, status reports, feasibility reports, manuals, etc,

The functions of reports are:

a) To provide information to the readers / users.

b) To initiate action.

c) To record for future use, or to maintain the history of a job or an organization for meeting legal requirements.

Development of a Format / standard for the preparation and presentation of information in a technical report has a number of advantages. Apart from increasing productivity and making activities habitual and automatic, a standard is helpful to the information specialist in a variety of ways:

- In the identification and collection of data / information to be included in the report.
- In the presentation of information in a technical report, it seeks to maintain uniformity and consistency.

It should, however, be emphasized that a format/ standard should be modified whenever a situation warrants. Normally, the components include: Title, Text, Conclusions, Recommendations, Acknowledgements, Bibliography, Tables, Figures / Illustrations, Appendices, Index, etc.

The text of the report should have the following sections /parts:

Introduction : Obviously, the introduction should provide the background to the study as also the work done earlier in regard to the problem being studied. The proposed study highlighting the need for study on the chosen subject; the statement of the problem; the specific objectives in relation to the problem, etc;

Methodology : Include description of Materials, Techniques, Equipments, Theoretical /Experimental/ Analytical methods, Assumptions, Limitations, etc. Provide all information necessary that would facilitate repeating and / or evaluating the study.

Data : discuss the pertinent data gathered, the key information of the report. Put supporting or background data not critical to the flow of the main report.

Analysis : provide an analysis drawn from the data presented. Note any deficiencies or unexplained data.

Conclusions : based on the analysis, draw conclusions relevant to the study.

Recommendations : suggest follow-up action.

Appendices : Present detailed supporting information that would detract from the flow of the report but is required as supplementary information.

Regarding the language and style, the suggestions given in the earlier chapters especially Chapter C are equally applicable in the context of preparation & presentation of information in Technical Reports. (Seetharama, 1997).

2.3.3. Progress Report

One writes a progress report to inform a supervisor, guide, associate, or customer about progress made on a project / thesis over a certain period of time. The project can be the design, construction or repair of something, the study or research of a problem or question, or gathering of information on a technical subject.

In the progress report, one explains any or all of the following:

- How much of the work is complete?
- What part of the work is currently in progress?
- What work remains tc be done?
- What problems or unexpected things, if any, have arisen?
- How the project is going in general?

Progress reports have several important functions:

1. Reassure recipients that you are making progress, that the project is going smoothly, and that it will be completed by the expected date.
2. Provide recipients with a brief look at some of the findings or some of the work of the project.
3. Give recipients a chance to evaluate the work on the project and suggest changes.
4. Give one a chance to discuss problems in the project and thus to forewarn recipients.
5. Force one to establish a work schedule so that the project will be completed on time.

Progress reports include the following parts:

1. An introduction that reviews the history of the project(s) beginning as well as the purpose and scope of the work.
2. A detailed description of the project.
3. An overall appraisal of the project to date, which usually acts as a conclusion.

Progress reports summarize

- Work accomplished in the preceding period(s)
- Work currently being performed
- Work planned for the next period(s)
- Individual tasks of the project

2.3.4. Review Article / State-of-the-art-report / Monographs

A review is a document which provides an overview of the developments that have taken place in a subject (s) during a certain time period, say the preceding year(s). While it was prepared earlier only by subject specialists, it is being prepared also by information professionals who have been trained in the techniques of information analysis and consolidation. The processes involved include selection, analysis, evaluation, restructuring and repackaging. The review document so prepared may be historical, descriptive, popular, critical, etc whose purpose is to keep users aware of developments taking place over a period of time in their respective areas of specialization.

The methodology to be followed in the preparation of TCs involving work in the three planes of work - -Idea, Verbal and Structural / Notational planes - - described in Chapter C is applicable in the preparation of a review article / state-of-the-art-report / monographs. It is emphasized here that the review article should be written in a language which involves the use of homonym-free ideas in a language. In this context, it is worthwhile mentioning that Ranganathan's *Prolegomena to Library Classification* would be very useful in review document preparation.

Monographs are similar to Review documents, generally prepared by subject specialists, and is a comprehensive, authoritative, analytical, exhaustive and evaluative written presentation.

2.3.5. Newsletters / House Journals / In-house Bulletins

Newsletters may be defined as a serial consisting of one or few pages, usually printed, and containing nascent information or news which is of interest to a special or particular group of users and published periodically. They present information in a precise and concise manner keeping in view the differential requirements of different categories of users. In other words, they are a medium of dissemination of news and activities relevant to the institution for the benefit of users and have a limited circulation.

The objectives & functions of Newsletters are:

- Serves as a type of publicity material which aims at announcing in capsule form information about current events, activities, publications, research studies, products, designs, employment news, etc of relevance to the users concerned.
- Informs users, especially, academic & research staff, about forthcoming conferences, seminars, symposia, meetings, etc to facilitate their participation.
- Publishes timely and relevant activities of current interest. For example: Topic of the Day.
- Serves as a medium for reporting and communicating information about social and cultural activities, personnel information -- appointments, transfers, retirements, marriages,

births, deaths, etc -- of the organization.

The contents of the Newsletters depend upon:

- Whether it is an announcement mechanism of meetings, workshops, forthcoming events to be held; or Scientific / Technical / Popular; or
- A combination of both.

The methodology outlined in Chapter C would be relevant here also and following are the steps:

- Identification of users and their informational requirements
- Scope definition
- Collection of information
- Appraisal of information collected for inclusion in the newsletter
- Presentation of data / information — textual and / or non-textual, format, etc.
- Finalisation
- Physical production of the Newsletter.

To conclude, it becomes clear that Newsletters are popular, non-technical medium of communication catering to different user groups. It aims at keeping users abreast of current happenings as well as future happenings. In its content, it can be best described as "Jack of all categories of information and master of none." However, a complacent attitude in its preparation should be avoided by the information professionals.

2.3.6. Scripts for Audio-Visual Presentations

A standard form of audio-visual presentation is a series of transparencies / slides in a revolving projector, wherein slides are synchronized with the sound track on a tape recorder. The slides are changed by a pulse recorded over the voice commentary on the tape. The writer's job is to write the supporting strip using a predetermined slide series. In this case, a simple brief on the nature of the subject may be enough to complete the writing of the script. For the A-V presentation, a person with a pleasant voice may be chosen to record the script. The entire approach in the production of the A-V presentation should be in the same way as any other writing task.

Some common features suggested by Evans (1983) flow from the following:

1) The material should be fully understandable and written to the level of expertise of the audience.
2) A general overview should be followed by a progressive break down into relevant facts and details.
3) For conceptual material, abstract ideas may be illustrated by anecdotes or concrete examples.
4) The A-V scripts are by nature highly concentrated and equally highly selective in terms of detailed material.
5) The length, a reasonable rule of thumb, is that 2000 words can be spoken in 15 minutes. At all events, it is advisable to read the script through at the projected pace with one eye on a stop watch

3. CONTENT DEVELOPMENT AND WEBPAGE DESIGN

Writing and preparing content for web requires a different approach from writing and preparing content for print documents and publications.

Further, writing content for the web that grabs attention and gets the results one wants requires special skills and attention.

Web content development is the process of researching, writing, gathering, organizing, and editing information for publication on websites.

The type of content one produces influences the popularity & profitability of your website. The content needs to be interactive and educational if one is operating a paid membership website. Further, content goes a long way in establishing the value of the website.

Hence, having the right content development strategy can make a big difference in how fast the site grows. Therefore, "Learn how to create and use the right content development strategy as a powerful means to achieve your website goals."

The following are Guidelines for Content Development Strategy (CDS):

1) Content needs to capture attention.
   - People prefer to read short articles
   - If you want many readers, focus on short and scannable content
   - If you want people who really need a solution, focus on comprehensive coverage

Jokob Nielsen suggests a CDS which mixes both long and short articles because it fulfils the needs of all. One way of doing this is to create content that is short and within it, include links to more detailed articles and in-depth information on the other pages, i.e. optimize CDS for your users' needs.

2) Content Development should not be separated from Strategic Planning.
   - Content should be considered as part of your overall master plan and not just an isolated necessity or discipline.
   - For generating revenue by volume in search engine traffics, "Sacrificing quality for quantity" is not desirable.
   - Strategy for "Content that makes you an expert", a different approach - - -content should demonstrate your knowledge and skills - - - needs to be adopted".

3) Content differentiation helps you survive and succeed in a crowded space.

- Content is only as good as the degree to which the reader finds it useful / entertaining.
- Subjectivity of content consumption is one of the main reasons why content dissemination and distribution channels are sometimes more important than the actual content quality.

(www.apdip.net/content)

Chapter E

EDITING PROCESS

1. WHAT IS EDITING?

In spite of various rules, criteria and guidelines, technical communications prepared will not be suitable for publication. The reasons may be either deviation from the criteria or omissions and in some cases errors. While elimination of errors is vital, there is more to editing than correcting mistakes. In other words, no writer on his own can produce a text that communicates precisely with the reader. Therefore, editing becomes essential. The question then is 'What is editing?". To edit simply means to prepare a document either for publication or presentation, by correcting, revising, or adapting. Thus, editing would consist of :

- Substantive editing (structural or text editing)
- Style editing (language & format editing)
- Copy editing (copy marking for the press)

In other words, the Editor is concerned with three aspects of a text:

- Does it conform to special or house style?
- Is the flow of material logical?
- Is the grammar and punctuation correct? (Evans, 1983)

Evans has given the following checklist of points to watch in editing:

a) Conformity to specification headings.
b) Page numbering.
c) Paragraph numbering.
d) Conformity of contents list to text.
e) Lay out of illustrations as per specification.
f) Unusual words or phrases.
g) Unexplained abbreviations.
h) Long words that may not be understood.
i) Unnecessary words -- adjectives, adverbs especially.
j) Undefined technical terms.
k) Balance of text -- length of paragraphs, etc.
l) Ambiguities in flow of text.
m) Punctuation.
n) Grammatical construction.
o) Spelling.
p) Conformity of meaning of text to what was intended.

On the other hand, Madhushree (2008) in her Blog mentions three levels of technical editing, namely:

i. Editing the text for grammar, spelling, and style errors.
ii. Editing the organization of the content, emphasis, appropriateness of content and adherence to standards and guidelines.
iii. Editing for accuracy of content, visual presentation, formatting issues, errors in the Table of Contents, and Indexes that have not been updated. She has given a detailed list of areas which editing should involve in her blog.

From the above, it is clear that editing is not an easy task and requires many literary-type skills as well as familiarity with the technical background of the subject matter, knowledge of legal aspects of publishing, administrative and marketing skills, etc.

Viewing from a slightly different perspective, editing proper consists of the function of elimination of errors apart from maintaining consistency and uniformity in style and presentation. Errors may pertain to format and / or content.

For eliminating errors of format, the editor should view the body of the document to ensure proper grammar and diction. For this, the following operations need to be done:

a) Correct all errors.

b) Singular subject with plural verbs or the reverse needs to be changed.

c) Split infinitives and dangling expressions.

d) Check spelling of words.

e) Check for correct punctuation.

On the other hand, invisible errors of content pose a serious problem. These are of two types, namely, omission of substantive material, and nomenclature. To avoid these errors, the editors should:

i. Develop the ability to find clues that indicate missing data / information.

ii. Ensure the proper and consistent use of nomenclatures in the document concerned.

In addition, the editor should categorise and cross reference the documents and prepare the necessary indexes.

Thus, the four major areas of difficulty in a text to which the editor should attend are:

1. Logical flaws in the organization of the content.
2. Poor paragraph development.
3. Complexities in sentence structure.
4. Incorrect language usage and vocabulary.

2. COPY EDITING

What is copy editing? It is the work that an editor does to improve the formatting, style, and accuracy (not of content) of a manuscript. It refers to written text for type setting, printing or publication. It means marking instructions on the manuscript for Composer, Printer, and Keyboard operator and thereby obtains consistency.

It has been said that the "Five Cs" summarise the copy editor's job. Make the copy:

- Clear
- Correct
- Concise
- Comprehensible and
- Consistent.

In other words, make it say what it means, and mean what it says. Typically, copy editing: Involves correcting spelling, punctuation, grammar, terminology / jargon, and semantics; Ensures that the typescript adheres to the publisher's house style; and adding headlines and standardized headers, footers, etc. The editor may also bridge a text by "cutting and trimming" it to reduce its length to fit publishing or broadcasting limits or to improve its meaning. (Wikipedia, Baskette et al, Butcher et al).

Skills for copy editors include:

- An excellent command of language
- General knowledge for spotting errors
- Good critical thinking skills in order to recognize inconsistencies
- Interpersonal skills for interaction with writers and other copy editors and a sense of style.

3. PROOF READING

Proof reading traditionally means reading a proof copy of a text in order to detect and correct any errors. In its turn, a proof copy is a version of manuscript that has been typeset after copy editing. Generally, it is done by the authors on special sheets provided by the printer. "Proofs" are example pages taken from the printing medium as a set or photographed and fall generally into three kinds:

- Galley proofs
- Page proofs
- Machine proofs

The first two categories of proofs are representative of the textual arrangement of the type, but not of the final print "finish". The machine proof, on the other hand, approximates the condition of the printed book in areas of quality. While galley proofs do not indicate the eventual configuration of the pages in the book, the page proof indicating actual book pages are taken when the lengths of run-on-text are composed into the correct page sizes.

It is common that proof typescripts often contain typographical errors due to mis-typing. Therefore, the proof reader checks the typeset copy and mark errors using standard proof readers' marks showing what are to be corrected. The International Standard ISO-5776 or The British Standard BS- 5261 and Bureau of Indian Standards 1250-1958 may be used for the purpose of indicating corrections to be carried out.

4. EDITORIAL TOOLS

By now, one would have realized that technical editing which includes structural or text editing, style editing (language and format editing), copy editing and proof reading, etc is not an easy task. Hence, the role of the editor becomes very important as his output will be scrutinized by the readers and peers of his profession. In this process of editing, the editor uses several tools like dictionaries, manuals, standards, specifications, thesauri, books of quotations, etc to check the veracity of facts, correct representation of forms, uniformity in style and adherence to standards. Some of these tools are discussed here.

4.1. Dictionaries

Dictionaries are reference books containing an alphabetical list of words, with information given for each word, usually including meaning(s),pronunciation, etymology, spelling(s), capitalization, different uses of words, abbreviation, italicization, different kinds of plural formation, authenticity, translation into other languages, etc.

Some definitions of dictionary on the web indicating their use are given below:

- A reference book containing an alphabetical list of words with information about them.
- A dictionary is a book or collection of words in a specific language, often listed alphabetically, with definitions, etymologies, pronunciations, etc.
- An alphabetical listing of words with their meanings, spellings, variant forms, pronunciation, etymology, synonyms, antonyms, usage examples in a single language.

Some dictionaries available in the market are:

1. Longman English Dictionary.
2. Concise Oxford Dictionary.
3. Merriam-Webster Online Dictionary.
4. The Australian Slang Dictionary.
5. Dictionary of Proverbs.
6. Dictionary of Quotations.
7. Dictionary of chemistry, etc.

4.2. Style Manuals

Style of writing is a function of a textual narrative. Its characteristics are: Clarity, Cadence and

Appropriateness. Bad style employs jargon, cliché, inappropriate ornament and robotic use of state imagery and outworn phrases.

A style guide or style manual is a set of standards for design and writing of documents, either for general use or for a specific publication or organization. Web style guides focus on a publication's visual and technical aspects, prose style, best usage, grammar, punctuation, spelling and fairness.

For purpose of style editing, the following Style Manuals would be useful:

1. Wikipedia: Manual of Style.
2. Chicago Manual of Style. Edn 15. University of Chicago.2003.
3. Style Manual: University of Minnesota.
4. Google INTERCAL style guide.
5. The New York Times Manual of Style and Usage.
6. The Elements of Style. William Strunk Jr.
7. Style guides of ISO.
8. A Manual for writers of term papers, theses & dissertations *by* Kate L. Turbian.
9. ACS Style Guide. (American Chemical Society).
10. American Medical Association Manual of Style.
11. Geosciences Reporting Guidelines.
12. IEEE Style.
13. MHRA Style Guide.
14. MLA Style Manual. Edn 3 and MLA Handbook for writers of research papers, Edn 7.
15. Scientific Style and Format: The CSE Manual for authors, editors, and Publishers. Edn 7.
16. The Style Manual for Political Science.

4.3. Standards

In the context of technical editing, standards play

an important role in relation to presentation of information in an appropriate manner. Two organizations, namely, International Standards Organization and Bureau of Indian Standards (BIS) have brought out many standards relevant to technical editing. They are, in fact, another set of powerful tools for technical editing. For example, BIS has published a number of Indian Standards for documentation pertaining to bibliographic references, books and periodicals, proof correction, typography, etc.

Some of them are listed below:

	Title	IS No.
1.	Bibliographical Reference: Recommendations for bibliographical reference.	2381-1963
2.	General structure of preliminary pages of a book	790 – 1956
3.	Guide of abbreviation of words in title of periodicals Using Roman Alphabet (first revision)	18 – 1970
4.	Guide for layout of learned periodicals	4 - 1963
5.	Guide for preparation of manuscript of an article in a Learned periodical	4731 – 1968
6.	Half-title leaf of a book	791 – 1956
7.	Practice for Table of Contents	791 – 1956
8.	Title page and back of title page (revised)	792 – 1964
9.	Practice for alphabetical arrangement	382 – 1952
10.	Rules for making Alphabetical Indexes	1275 – 1958
11.	Specification for Library Catalogue & Abstract card	7150 – 1974
12.	Proof corrections for printers & authors	1250 – 1958

Epilogue

"A common criticism of Technical Writing is pomposity - - ponderous, dull, and unreadable. Technical Writing can be too abstract, too passive stylistically, and downright cautious".

The question then is: Why do technical writers write badly? Possible reasons include:

- They write to a format well entrenched and slavishly adhered to.
- They want completeness and precision, regardless of the cost to clear and straightforward communication. Technical details, to the point of minutia, are more likely to confuse than inform the reader.
- Brevity or terseness is more desirable than clarity.
- They want to impress with the extent of vocabulary or the ability to construct complicated sentences. Watch out for those who write with their motto, "Never use simple words when a complicated phrase will do!" (CPE 632: Guide for Technical Writing.)

To avoid this criticism, one needs to follow the principles & recommended practices outlined above.

Perhaps, the Library & Information Science Profession should be interested in Technical Writing and view it as an alternative career option as well as to be successful librarians and information professionals.

To conclude, Good Technical Writing ensures conformity to Ranganathan's Five Laws of Library Science.

Bibliographical References

1. BASKETTE (FK) et al. The Art of Editing.
2. BUTCHER (J) et al. Butcher's copy editing. (*In* The Cambridge Handbook for Editors, Copy Editors and Proof Readers. Edn 4. Cambridge University Press. Cambridge. 2006)
3. CAMBELL (WG) and BALLOU (SV). Form and style. 1978 Houghton Mifferi co., Boston.
4. CHANDRA (A) and SAXENA (P). Style manual for writing theses, dissertations and papers in Social Sciences. 1979. Metropolitan Book Pvt Ltd. Delhi.
5. Copy Editing. (*In* Wikipedia, the free encyclopedia. *en.wikipedia.org/ wiki/ copyediting*
6. ELRIDGE (J C). By the book: How to develop a Company Style Manual. 2002. Tetra Tech Inc, Santa Barbara, USA.
7. EVANS (J). Beginner's Guide to Technical Writing. Butterworth & Co Ltd.1993.
8. GOPALAKRISHNAN (NK), Linguistic approach to condensation of information in Natural Languages. (DRTC Annual Seminar 18; 1981; Paper BE).
9. GOPINATH (MA). Typology for structural relations for Information Analysis and Consolidation

processes. (Library Science. 29; 1992; 139 -142.)

10. GRAVEY (WD). Communication: The essence of science facilitating information exchange among libraries, scientists, engineers and students. Oxford Pergamon Press. 1979.
11. KASTENS (K) et al. Comp. How to write your thesis.
12. LAMPE (William L). Technical Writing Overview. 1970
13. MORRISON (M A). Tips on scientific writing. 2004.
14. McMURREY (David A): Online technical writing: Text book. https://www.tu_chemintz.de/phil/english/chairs/linguist/.../acctoc.html
15. NEELAMEGHAN (A). Presentation of ideas in technical writing. Edn 1. 1975.
16. NEELAMEGHAN (A) and SEETHARAMA (S). Reference Framework: An Aid to Information Analysis and Consolidation. (DRTC Annual Seminar. 18; 1981; Paper BD).
17. Online Technical Writing: progress reports. *www.io.com/-heexres/textbook/progrep*
18. PARKER (C). Using and presenting information. (In Bradfield (VJ), Ed. Information Sources in Architecture. 1983. Butterworths. P 76 – 95.)
19. RAJAN (Rajashri). Technical Writing: 'Write' Away! (*In* Ramesh Babu (B) and others, Ed. Information Management: Trends and Issues.2003. Researchco Book Centre, New Delhi.)
20. RANGANATHAN (SR). Prolegomena to Library classification. Edn 3. 1967. Asia Publishing House. Bombay.
21. Readability and reading ages of school science text books. *www.timetabler.com/reading.html*
22. SARACEVIC (T) and WOOD (JS). Consolidation of Information: A handbook of evaluation,

restructuring and repackaging of scientific and technical information. 1981. UNESCO. Paris.

23. SEETHARAMA (S). Guidelines for Abstracting. (Paper presented at the IASLIC Seminar, Madras. 1978).
24. SEETHARAMA (S). Information Consolidation and Repackaging: Framework, Methodology and Planning. 1997. Ess Ess Publications. New Delhi.
25. SEETHARAMA (S). Technical Writing: Principles and Strategies. (National Workshop on Technical Writing. Feb 2005. Dept. of Library & Information Science, University of Kerala. Trivandrum.)
26. TURNER (BT). Effective Technical Writing and Speaking. 1974. Business Books Limited, London.
27. WEBER (J H). Developing a Departmental Style Guide. 2011. http://techwhirl.com/developing-a-departmental-style-guide/
28. What is Copy Editing? Activities and 5 Cs.Freelance Panda.com, 2008, Webpage:Free P-Services
29. Wikepedia, the free encyclopedia. *En.wikipedia.org/wiki/readability.*
30. WILSON (JR). Some Guidelines On Technical Writing jwilson@eos.ncsu.edu

INDEX

AUTHOR INDEX

Note: Each entry consists of the name of the author followed by Page Number

SUBJECT INDEX

Note: Each entry consists of the name of the subject or kernel term followed by Page Number(s).

def = definition; *irt* = *in relation to; TC* = *Technical communication; TW* = *Technical writing*

Appropriateness. Bad style employs jargon, cliché, inappropriate ornament and robotic use of state imagery and outworn phrases.

A style guide or style manual is a set of standards for design and writing of documents, either for general use or for a specific publication or organization. Web style guides focus on a publication's visual and technical aspects, prose style, best usage, grammar, punctuation, spelling and fairness.

For purpose of style editing, the following Style Manuals would be useful:

1. Wikipedia: Manual of Style.
2. Chicago Manual of Style. Edn 15. University of Chicago.2003.
3. Style Manual: University of Minnesota.
4. Google INTERCAL style guide.
5. The New York Times Manual of Style and Usage.
6. The Elements of Style. William Strunk Jr.
7. Style guides of ISO.
8. A Manual for writers of term papers, theses & dissertations *by* Kate L. Turbian.
9. ACS Style Guide. (American Chemical Society).
10. American Medical Association Manual of Style.
11. Geosciences Reporting Guidelines.
12. IEEE Style.
13. MHRA Style Guide.
14. MLA Style Manual. Edn 3 and MLA Handbook for writers of research papers, Edn 7.
15. Scientific Style and Format: The CSE Manual for authors, editors, and Publishers. Edn 7.
16. The Style Manual for Political Science.

4.3. Standards

In the context of technical editing, standards play

correct representation of forms, uniformity in style and adherence to standards. Some of these tools are discussed here.

4.1. Dictionaries

Dictionaries are reference books containing an alphabetical list of words, with information given for each word, usually including meaning(s),pronunciation, etymology, spelling(s), capitalization, different uses of words, abbreviation, italicization, different kinds of plural formation, authenticity, translation into other languages, etc.

Some definitions of dictionary on the web indicating their use are given below:

- A reference book containing an alphabetical list of words with information about them.
- A dictionary is a book or collection of words in a specific language, often listed alphabetically, with definitions, etymologies, pronunciations, etc.
- An alphabetical listing of words with their meanings, spellings, variant forms, pronunciation, etymology, synonyms, antonyms, usage examples in a single language.

Some dictionaries available in the market are:

1. Longman English Dictionary.
2. Concise Oxford Dictionary.
3. Merriam-Webster Online Dictionary.
4. The Australian Slang Dictionary.
5. Dictionary of Proverbs.
6. Dictionary of Quotations.
7. Dictionary of chemistry, etc.

4.2. Style Manuals

Style of writing is a function of a textual narrative. Its characteristics are: Clarity, Cadence and